Mathematics Research Developments Series

BOUNDARY PROPERTIES AND APPLICATIONS OF THE DIFFERENTIATED POISSON INTEGRAL FOR DIFFERENT DOMAINS

MATHEMATICS RESEARCH DEVELOPMENTS SERIES

Quasi-Invariant and Pseudo-Differentiable Measures in Banach Spaces
Sergey V. Ludkovsky
2009. ISBN 978-1-60692-734-2

Boundary Properties and Applications of the Differentiated Poisson Integral for Different Domains
Sergo Topuria
2009. ISBN 978-1-60692-704-5

Mathematics Research Developments Series

BOUNDARY PROPERTIES AND APPLICATIONS OF THE DIFFERENTIATED POISSON INTEGRAL FOR DIFFERENT DOMAINS

SERGO TOPURIA

Nova Science Publishers, Inc.
New York

For permission to use material from this book please contact us:
Telephone 631-231-7269; Fax 631-231-8175
Web Site: http://www.novapublishers.com

LIBRARY OF CONGRESS CATALOGING-IN-PUBLICATION DATA
Available upon request

ISBN 978-1-60692-704-5

Published by Nova Science Publishers, Inc. ✢ New York

Contents

Preface

The boundary properties of derivatives of the Poisson integral for a circle were investigated in full detail by P. Fatou ([33]). In particular he proved

Theorem A. *If there exists a finite derivative $f(x_0)$, then the derivative of the Poisson integral has $f'(x_0)$ as its nontangential limit* for a function f.

Theorem B. *If there exists a finite or an infinite symmetric derivative of first order $f^*_{(1)}(x_0)$, then the derivative of the Poisson integral has $f^*_{(1)}(x_0)$ as its radial limit* for a function f.

In this book we investigate the boundary properties of the differentiated Poison integral for various domains such as circle, ball, half-plane, half-space, bicylinder and give the application of these properties for the solution of the Dirichlet problem when the boundary function is measurable and finite almost everywhere.

Chapter I deals with the boundary properties of derivatives of any order of the Poisson integral for a half-plane when the integral density has a generalized derivative (in this sense or another) of any order. The boundary properties of a first order derivative of the Poisson integral for a half-plane, when the integral density has an ordinary derivative, were investigated by A.G. Jvarsheishvili in [13]. In Section 1.3 it is shown that the existence of a finite symmetric derivative of the density integral does not ensure the existence of an angular limit of a Poisson integral derivative. In Section 1.4, the Dirichlet problem is solved in N.N. Luzin's formulation for a half-plane.

Chapter II investigates the boundary behavior of derivatives of any order of the Poisson integral for a circle. It is shown in Section 2.3 that Theorem B cannot be made stronger in terms of the existence of an angular limit. Various analogues of the Fatou theorems are proved for generalized derivatives of any order and the fact that they cannot be strengthened in a certain sense is proved.

In Chapter III consideration is given to the problems connected with the boundary properties of derivatives of the Poisson integral for a ball (of any finite dimension), where the differentiation operator is a Laplace operator on the unit sphere, i.e., an angular portion of the Laplace operator written in terms of spherical coordinates.

For $k = 3$, the boundary properties of the first and second order partial derivatives of the Poisson integral for a ball have been studied by O.P. Dzagnidze in [18–25].

In Section 3.3, we introduce the notions of generalized Laplace operators on the unit sphere, while in Section 3.5 we prove the theorems on the boundary properties of an integral $D_k^r U(f;\rho,\theta_1,\theta_2,\dots,\theta_{k-2},\varphi)$ $(k>2,\ r\in N)$, where $U(f;\rho,\theta_1,\theta_2,\dots,\theta_{k-2},\varphi)$ is the Poisson integral for the unit ball in R^k $k>2$, and D_k is a Laplace operator on the unit sphere S^{k-1}. In this section it is proved that the obtained results are non-improvable in a certain definite sense. In Section 3.6 the Dirichlet problem is solved for a ball when the boundary function is measurable and finite almost everywhere.

Chapter IV deals with the boundary properties of partial derivatives and differentials of arbitrary order of the Poisson integral for a half-space R_+^{k+1} $(k>1)$. In Section 4.2 the existence of a continuous function with first order finite partial derivatives at the point $x^0=(x_1^0,x_2^0,\dots,x_k^0)$ is established, while first order partial derivatives of the Poisson integral of this function have no boundary values at the same point even with respect to the normal. Hence there arises a question how to generalize the notion of derivatives of functions of several variables so that Fatou type theorems be valid for the Poisson integral $U(f;x,x_k+1)$. In Sections 4.1, 4.3, 4.6, 4.8 and 4.10 we introduce the notions of a generalized partial derivative, a generalized differential and a generalized spherical derivative for functions of several variables. In Sections 4.2, 4.4, 4.7, 4.9 and 4.11 we prove Fatou type theorems on the boundary properties of partial derivatives and differentials of arbitrary order for the Poisson integral in the case of a half-space, when the integral density has a generalized partial derivative, a generalized differential or a generalized spherical derivative. The results obtained in this chapter show that in the case of a half-space the boundary properties of derivatives of the Poisson integral depend essentially on on a form in which the integral density is differentiable. It is also proved that the obtained results are non-improvable in a certain sense. In Section 4.5 the Dirichlet problem is solved for a half-space when the boundary function is measurable and finite almost everywhere.

Chapter V deals with the boundary properties of the differentiated Poisson integral for a bicylinder. It is proved that whatever the smoothness of the Poisson integral density is in the neighborhood of a given point, it does not ensure the existence of boundary values of partial derivatives of the Poisson integral at the considered point. Furthermore, the sufficient conditions are found for the existence of limiting values of first and second order partial derivatives of the Poisson integral in the case of a bicylinder. It is proved that the obtained results cannot be strengthened (in a certain sense). In Sections 5.3 and 5.5 we consider the problem os representing an arbitrary measurable and almost everywhere finite function of two variables by a double trigonometric series both in the case of spherical convergence and in the case of convergence in the Pringsheim sense.

Acknowledgement. The author expresses his sincere gratitude to Prof. O.P. Dzagnidze and Prof. G. E. Tkebuchava who read the manuscript and made a number of useful comments.

The present monograph is devoted to the investigation of boundary properties of the differentiated Poisson integral. It is proved that the boundary properties of the differentiated Poisson integral for different types of domains (a circle, sphere, half-plane, half-space, bicylinder) differ significantly from each other and depend on a sense in which the integral density is differentiable. The theorems proven here are, in a definite sense, unimprovable. On the basis of the obtained results, the Dirichlet problem for a sphere and half-space (of any finite dimension) is solved in case the boundary function is only measurable and finite almost everywhere.

The monograph is intended for scientific workers, persons working for Doctor's degree, postgraduates and students interested in the function theory.

Editor: Prof. T. I. Akhobadze

Reviewrs: Prof. O. P. Dzagnidze
Prof. Sh. T. Tetunashvili

Chapter 1

Boundary Properties of Derivatives of the Poisson Integral for a Half-Plane

1.1 Notation, Definitions and the Well-Known Statements

We use the following notation: $R = R^1 =]-\infty;\infty[$; $\widetilde{L}(R)$ is a set of measurable functions $f(x)$ such that

$$\frac{f(x)}{1+x^2} \in L(R); \quad R^2_+ = \{(x,y) \in R^2; \quad y > 0\};$$

$U(f;x,y)$ is the Poisson integral of the function $f(x)$ for a half-plane R^2_+, i.e.,

$$U(f;x,y) = \frac{y}{\pi}\int\limits_R \frac{f(t)dt}{(t-x)^2+y^2} = \frac{1}{\pi}\int\limits_{-\infty}^{\infty} P(t-x,y)f(t)dt,$$

$$P(t-x,y) = \frac{y}{(t-x)^2+y^2}.$$

The symbol $M(x,y) \stackrel{\wedge}{\longrightarrow} P(x_0,0)$ means that a point $M(x,y) \in R^2_+$ tends to $P(x_0,0)$ along the nontangential path, i.e., there exists a positive number C such that

$$\frac{|x-x_0|}{y} < C.$$

$M(x,y) \to P(x_0,0)$ means that the point $M(x,y)$ tends to $P(x_0,0)$ in an arbitrary manner, remaining in the half-plane R^2_+.

Assume (see [35], p.92) that the function $f(x)$ is defined in some neighborhood of the point x_0 and that there exist constants $\alpha_0, \alpha_1, \dots, \alpha_r$ such that for small t,

$$f(x_0+t) = \alpha_0 + \alpha_1 t + \dots + \alpha_{r-1}\frac{t^{r-1}}{(r-1)!} + [\alpha_r + \varepsilon(t)]\frac{t^r}{r!},$$

where $\varepsilon(t) \to 0$ as $t \to 0$. Then that the function $f(x)$ is said to have the generalized rth derivative $f_{(r)}(x_0)$ at the point x_0 and, by definition, $f_{(r)}(x_0) = \alpha_r$. It is clear that $\alpha_0 = f(x_0)$, $\alpha_1 = f'(x_0)$. Moreover, if there exists $f^{(r)}(x_0)$, then there likewise exist $f_{(r)}(x_0)$ and $f^{(r)}(x_0) = f_{(r)}(x_0)$. The converse statement is not true. This definition is due to Peano.

Let us now recall the notion of a generalized symmetric derivative which belongs to Valée-Poussin.

Let r be an odd number. If there exist constants $\beta_1, \beta_3, \dots, \beta_r$ such that

$$\frac{f(x_0+t) - f(x_0-t)}{2} = \beta_1 t + \beta_3\frac{t^3}{3!} + \dots + \beta_{r-2}\frac{t^{r-2}}{(r-2)!} + [\beta_r + \varepsilon(t)]\frac{t^r}{r!},$$

where $\varepsilon(t) \to 0$ as $t \to 0$, then β_r is called the r-th generalized symmetric derivative or, shortly, the r-th symmetric derivative of f at the point x_0. We denote this derivative by the symbol $f^*_{(r)}(x_0)$, i.e., $\beta_r = f^*_{(r)}(x_0)$. The same definition is given for even r, but the difference $f(x_0+t) - f(x_0-t)$ should be replaced by the sum $f(x_0+t) + f(x_0-t)$.

Clearly,

$$\beta_1 = \lim_{t\to 0}\frac{f(x_0+t) - f(x_0-t)}{2t} = f^*_{(1)}(x_0), \tag{1.1}$$
$$\beta_0 = f(x_0), \quad \beta_2 = \lim_{t\to 0}\frac{f(x_0+t) + f(x_0-t) - 2f(x_0)}{t^2} = f^*_{(2)}(x_0).$$

It can be easily verified that the existence of derivatives $f_{(r)}(x_0)$ implies the existence of derivatives $f^*_{(r)}(x_0)$ and their equality (see [35], p.93).

For symmetric derivatives, from the existence of $f^*_{(r)}(x_0)$ follows the existence of $f^*_{(r-2)}(x_0)$, but not necessarily $f^*_{(r-1)}(x_0)$ (see [35], p.93).

If there exist functions $\alpha_i(x)$, $i = 0, 1, 2, \dots, r-1$, defined in the neighborhood of the point x_0 and a number α_r such that there exist limits $\lim\limits_{x\to x_0}\alpha_i(x) = \alpha_i$, and in the neighborhood of the point x_0 we have for small t,

$$f(x+t) = \alpha_0(x) + \alpha_1(x)t + \dots + \alpha_{r-1}(x)\frac{t^{r-1}}{(r-1)!} + [\alpha_r + \varepsilon(x,t)]\frac{t^r}{r!}, \tag{1.2}$$

where $\lim\limits_{(x,t)\to(x_0,0)}\varepsilon(x,t) = 0$, then we say that the function f has the generalized r-th derivative in a strong sense at the point x_0 and we define $\overline{f}_{(r)}(x_0) = \alpha_r$.

Clearly, $\alpha_0(x) = f(x)$, while

$$\alpha_1 = \overline{f}_1(x_0) = \lim_{(x,t)\to(x_0,0)} \frac{f(x+t)-f(x)}{t}. \tag{1.3}$$

From the above definition it follows that if $\overline{f}_{(r)}(x_0)$ exists, then there also exists $f_{(r)}(x_0)$, and they are equal.

Let us now give the definition of a generalized symmetric derivative in a strong sense. Let r be an even number. If there exist functions $\beta_0(x), \beta_2(x), \ldots, \beta_{r-2}(x)$ defined in the neighborhood of the point x_0, and a number β_r such that there exist limits $\lim\limits_{x\to x_0} \beta_{2i}(x) = \beta_{2i}$, $i = 0, 1, 2, \ldots, \frac{r-2}{2}$, and in the neighborhood of the point x_0 we have for small t:

$$\frac{(f(x+t)+f(x-t)}{2} = \beta_0(x) + \beta_2(x)\frac{t^2}{2!} + \cdots + \beta_{r-2}(x)\frac{t^{r-2}}{(r-2)!} + [\beta_r + \varepsilon(x,t)]\frac{t^r}{r!},$$

where $\lim\limits_{(x,t)\to(x_0,0)} \varepsilon(x,t) = 0$, then we say that f has the r-th generalized symmetric derivative in a strong sense at the point x_0, and we define $\overline{f}^*_{(r)}(x_0) = \beta_r$.

The same definition is given for an odd r, but the sum $f(x+t)+f(x-t)$ should be replaced by the difference $f(x+t)-f(x-t)$.

It is not difficult to see that from the existence of a derivative $\overline{f}_{(r)}(x_0)$ follows the existence of a derivative $\overline{f}^*_{(r)}(x_0)$ and their equality.

Clearly, $\beta_0(x) = f(x)$,

$$\beta_1 = \lim_{(x,t)\to(x_0,0)} \frac{f(x+t)-f(x-t)}{2t} = \overline{f}^*_{(1)}(x_0),$$

$$\beta_2 = \lim_{(x,t)\to(x_0,0)} \frac{f(x+t)+f(x-t)-2f(x)}{t^2} = \overline{f}^*_{(2)}(x_0).$$

The following statements are valid:

(1) The existence of $f'(x_0)$ implies the existence of $f^*_{(1)}(x_0)$ and $f^*_{(1)}(x_0) = f'(x_0)$. The converse statement is not true (see, for e.g., [115], p. 614).

(2) The existence of $f''(x_0)$ implies the existence of $f^*_{(2)}(x_0)$ and $f^*_{(2)}(x_0) = f''(x_0)$.

Using the function

$$f(x) = \int_0^x t\sin\frac{1}{t}dt$$

as an example it is easy to show that the converse statement is not true.

(3) The existence of $\overline{f}_{(r)}(x_0)$ implies the existence of $\overline{f}_{(i)}(x_0)$ and $f_i(x_0) = \alpha_i$ $(i = 1, 2, \ldots, r-1)$.

Indeed, let there exist $\overline{f}_{(r)}(x_0)$, i.e., let the equality (1.2) hold. Then

$$f(x+t) = \alpha_0(x) + \alpha_1(x)t + \cdots + \alpha_{r-2}(x)\frac{t^{r-2}}{(r-2)!} + [\alpha_{r-1} + \varepsilon_0(x)]\frac{t^{r-1}}{(r-1)!} + [\overline{f}_{(r)}(x_0) + \varepsilon(x,t)]\frac{t^r}{r!}, \tag{1.4}$$

$\varepsilon_0(x) \to 0$ as $x \to x_0$. Thus (1.4) can be rewritten as follows:

$$f(x+t) = \alpha_0(x) + \alpha_1(x)t + \cdots + \alpha_{r-2}(x)\frac{t^{r-2}}{(r-2)!} + [\alpha_{r-1} + \varepsilon_1(x,t)]\frac{t^{r-1}}{(r-1)!},$$

where $\varepsilon_1(x,t) = \varepsilon_0(x) + \frac{t}{r}\overline{f}_{(r)}(x_0) + \frac{t}{r}\varepsilon(x,t)$. which tends to zero as $(x,t) \to (x_0,0)$. The existence of the remaining derivatives can be shown analogously.

(4) The existence of $\overline{f}_{(1)}(x_0)$ implies the existence of $f'(x_0)$ and $\overline{f}_{(1)}(x_0) = f'(x_0)$. Using the function

$$f(x) = \begin{cases} x^2 \sin\frac{1}{x}, & \text{for } x \neq 0, \\ 0, & \text{for } x = 0 \end{cases}$$

as an example it is easy to show that the converse statement is invalid.

(5) If there exists $\overline{f}_{(1)}(x_0)$, then almost everywhere in the neighborhood of the point x_0 there exists $f'(x)$.

(6) If $f'(x)$ exists in the neighborhood of the point x_0 and is continuous at the point x_0, then $\overline{f}_{(1)}(x_0)$ exists too, and $\overline{f}_{(1)}(x_0) = f'(x_0)$.

The converse statement is not true: there exists a function $f(x)$ which at the point 0 has $\overline{f}_{(1)}(o)$, but in the neighborhood of the point 0 there exists almost everywhere a dense set whose every point does not contain $f'(x)$.

Indeed, let $\{r_k\}$ be a sequence of rational numbers, everywhere dense in the neighborhood of the point 0. Consider the function

$$\varphi(x) = \sum_{k=1}^{\infty} \frac{|x - r_k|}{2^k}.$$

$\varphi(x)$ is continuous and differentiable everywhere except for the points $r_1, r_2, \ldots$.

Let

$$f(x) = x\varphi(x).$$

It is clear that the function $f(x)$ is also differentiable everywhere except for the points $r_1, r_2, \ldots$, and $f'(o) = 0$. It is not difficult to show that $\overline{f}_{(1)}(o)$ also exists, and $\overline{f}_{(1)}(o) = 0$.

(7) If the derivative $f''(x)$ exists in the neighborhood of the point x_0 and is continuous at the point x_0, then there exists $\overline{f}_{(2)}(x_0)$, and $\overline{f}_{(2)}(x_0) = f''(x_0)$.

Note that the continuity of the derivative $f''(x)$ at the point x_0 is only the sufficient condition for the existence of $\overline{f}_{(2)}(x_0)$.

In the sequel it will be assumed that $f(x) \in \widetilde{L}(R)$.

1.2 Auxiliary Statements

The following lemma is valid.

Lemma 1.2.1. *For every $r \in N$ and $(x,y) \in R^2_+$, the following statements are valid.*

1) $I_i^{(r)} = \int\limits_{-\infty}^{\infty} \frac{\partial^r P(t-x,y)}{\partial x^r} t^i dt = 0,\ i = 0,1,2,\ldots,r-1;$

2) $I_r = \frac{1}{\pi} \int\limits_{-\infty}^{\infty} \frac{\partial^r P(t-x,y)}{\partial x^r} \frac{t^r}{r!} dt = 1;$

3) $\int\limits_{-\infty}^{\infty} \left| \frac{\partial^r P(t,y)}{\partial t^r} \right| |t|^r dt < C;$*

4) $\int\limits_{-\infty}^{\infty} \left| \frac{\partial^r P(t-x,y)}{\partial t^r} \right| |t|^r dt < C$ *for* $\frac{y}{|x|} \geq C > 0;$

5) $\sup\limits_{|t| \geq \delta > 0} \left| \frac{\partial^r P(t,y)}{\partial t^r} \right| (t^2+y^2)|t|^v < Cy,\ v = \overline{0,r}.$

Proof. Statement (1) is proved by induction. When $r = 1$ and $i = 0$, the validity of the statement follows from the equality

$$\int\limits_{-\infty}^{\infty} P(t-x,y)dt = \int\limits_{-\infty}^{\infty} P(t,y)dt = y \int\limits_{-\infty}^{\infty} \frac{dt}{t^2+y^2} = \pi.$$

For $r = 2$, we have

$$I_0^{(2)} = \frac{\partial^2}{\partial x^2} \int\limits_{-\infty}^{\infty} P(t-x,y)dt = 0$$

and

$$I_1^{(2)} = \int\limits_{-\infty}^{\infty} \frac{\partial^2 P(t-x,y)}{\partial x^2} tdt = \int\limits_{-\infty}^{\infty} \frac{\partial^2 P(t,y)}{\partial t^2} (t+x)dt$$

$$= \int\limits_{-\infty}^{\infty} \frac{\partial^2 P(t,y)}{\partial t^2} tdt + x \int\limits_{-\infty}^{\infty} \frac{\partial^2 P(t,y)}{\partial t^2} dt = \int\limits_{-\infty}^{\infty} \frac{3t^2-y^2}{(t^2+y^2)^3} tdt = 0.$$

Let us now assume that the equality

$$I_1^{(n)} = \int\limits_{-\infty}^{\infty} \frac{\partial^n P(t-x,y)}{\partial x^n} t^i dt = 0, \quad i = \overline{0,n-1}$$

*Here and in the sequel, by C we denote the absolute positive constants which may, generally speaking, be different in various correlations.

is fulfilled for $r = n$ and show that

$$I_1^{(n+1)} = \int_{-\infty}^{\infty} \frac{\partial^{n+1} P(t-x,y)}{\partial x^{n+1}}\, t^i dt = 0, \quad i = \overline{0,n}.$$

Indeed, the integration by parts yields

$$0 = I_i^{(n)} = (-1)^n \int_{-\infty}^{\infty} \frac{\partial^n P(t-x,y)}{\partial t^n}\, t^i dt$$

$$= \frac{(-1)^{n+1}}{i+1} \int_{-\infty}^{\infty} \frac{\partial^{n+1} P(t-x,y)}{\partial t^{n+1}}\, t^{i+1} dt = \frac{1}{i+1} \int_{-\infty}^{\infty} \frac{\partial^{n+1} P(t-x,y)}{\partial x^{n+1}}\, t^{i+1} dt.$$

Thus $I_i^{n+1} = 0$ when $i = \overline{1,n}$, and for $i = 0$ we likewise have

$$I_0^{(n+1)} = \frac{\partial^{n+1}}{\partial x^{n+1}} \int_{-\infty}^{\infty} P(t-x,y) dt = 0.$$

The validity of Statement (2) is also proved by induction. For $r = 1$, we have

$$I_1 = \frac{1}{\pi} \int_{-\infty}^{\infty} \frac{\partial P(t-x,y)}{\partial x}\, t dt = \frac{2y}{\pi} \int_{-\infty}^{\infty} \frac{(t-x)t}{[(t-x)^2+y^2]^2}\, dt$$

$$= \frac{2y}{\pi} \int_{-\infty}^{\infty} \frac{t(t+x)}{(t^2+y^2)^2}\, dt = \frac{2}{\pi} \int_{-\infty}^{\infty} \frac{t^2 dt}{(1+t^2)^2} = 1.$$

(see [10], p. 79).

When $r = 2$, we have

$$I_2 = \frac{1}{\pi} \int_{-\infty}^{\infty} \frac{\partial^2 P(t-x,y)}{\partial x^2} \frac{t^2}{2} dt = \frac{2y}{\pi} \int_{-\infty}^{\infty} \frac{3(t-x)^2 - y^2}{[(t-x)^2+y^2]^3}\, t^2 dt$$

$$= \frac{2y}{\pi} \int_0^{\infty} \frac{(3t^2-1)t^2}{(1+t^2)^3}\, dt = 1.$$

Assume that for $r = n$ the equality

$$I_n = \frac{1}{\pi} \int_{-\infty}^{\infty} \frac{\partial^n P(t-x,y)}{\partial x^n} \cdot \frac{t^n}{n!}\, dt = 1$$

is fulfilled.

Taking this into account, we can show that

$$I_{n+1} = \frac{1}{\pi} \int_{-\infty}^{\infty} \frac{\partial^{n+1} P(t-x,y)}{\partial x^{n+1}} \cdot \frac{t^{n+1}}{(n+1)!}\, dt = 1. \tag{2.1}$$

Indeed, using the integration by parts, we obtain

$$\int_{-\infty}^{\infty} \frac{\partial^{n} P(t-x,y)}{\partial x^{n}} \cdot \frac{t^{n}}{n!}\, dt = \int_{-\infty}^{\infty} \frac{\partial^{n+1} P(t-x,y)}{\partial x^{n+1}} \cdot \frac{t^{n+1}}{(n+1)!}\, dt,$$

which shows that the equality (2.1) is valid.

Let us now prove the validity of Statement (3). We have

$$|t|^r \frac{\partial^r P(t,y)}{\partial t^r} = |t|^r y \frac{\partial^r}{\partial t^r}\Big(\frac{1}{t^2+y^2}\Big) = |t|^r y \frac{I(t,y)}{(t^2+y^2)^{r+1}},$$

where $I(t,y)$ is a homogeneous polynomial of degree r of (t,y). Thus

$$\int_{-\infty}^{\infty} \Big|\frac{\partial^r P(t,y)}{\partial t^r}\Big| |t|^r dt = y \int_{-\infty}^{\infty} \Big|\frac{I(t,y)t^r}{(t^2+y^2)^{r+1}}\Big|\, dt \le C \int_{0}^{\infty} \frac{\sum_{v=0}^{r} t^{v+r}}{(1+t^2)^{r+1}} dt = O(1).$$

Statement (4) follows from Statement (3) if we take into account the conditions

$$\frac{y}{|x|} \ge \delta > 0.$$

Indeed, using the inequality $|a+b|^p \le 2^p(|a|^p + |b|^p)$ for $p \ge 1$, we obtain

$$\begin{aligned}
\int_{-\infty}^{\infty} \Big|\frac{\partial^r P(t-x,y)}{\partial t^r}\Big| |t|^r dt &= \int_{-\infty}^{\infty} \Big|\frac{\partial^r P(t,y)}{\partial t^r}\Big| |t+x|^r dt \\
&\le 2^r \int_{-\infty}^{\infty} \Big|\frac{\partial^r P(t,y)}{\partial t^r}\Big| |t|^r dt + 2^r |x|^r \int_{-\infty}^{\infty} \Big|\frac{\partial^r P(t,y)}{\partial t^r}\Big| dt \\
&= O(1) + C|x|^r \int_{-\infty}^{\infty} \frac{y|I(t,y)|}{(t^2+y^2)^{r+1}}\, dt \\
&= O(1) + \frac{C|x|^r y^{2+r}}{y^{2r+2}} \int_{-\infty}^{\infty} \frac{|I(\rho,1)|}{(1+\rho^2)^{r+1}}\, d\rho \\
&\le O(1) + C \int_{0}^{\infty} \frac{\sum_{v=0}^{r} \rho^v}{(1+\rho^2)^{r+1}} d\rho = O(1).
\end{aligned}$$

Statement (5) follows from the inequality

$$\Big|\frac{\partial^r P(t,y)}{\partial t^r}\Big|(t^2+y^2)|t|^v = y\frac{|I(t,y)|}{(t^2+y^2)^{r+1}}(t^2+y^2)|t^v|$$
$$\leq y\frac{|I(t,y)|}{(t^2+y^2)^{r+1}}(t^2+y^2)^{\frac{v}{2}+1} = y\frac{|I(t,y)|}{(t^2+y^2)^{r-\frac{v}{2}}}(t^2+y^2)^{r-\frac{v}{2}} < Cy,$$
$$v=\overline{0,r}, \quad |t|\geq\delta,$$

since $I(t,y)$ is a homogeneous polynomial of degree r. □

1.3 The Boundary Properties of Derivatives of the Poisson Integral for a Half-Plane

The boundary properties of a first order derivative of the Poisson integral for a half-plane in case the integral density has a finite ordinary derivative, are studied in [13]. The same problem is investigated in [91] for the case, where the integral density has a first symmetric derivative. In the same work, it is shown that the obtained results cannot be strengthened in the sense of the existence of an angular limit.

In this section, we study the boundary properties of derivatives of any order of the Poisson integral for a half-plane, when the integral density has a generalized derivative of arbitrary order.

The following theorems are valid (see [13], [91], [98]).

Theorem 1.3.1. (a) *If at the point x_0 there exists a finite $f^*_{(r)}(x_0)$, then*

$$\lim_{y\to 0+}\frac{\partial^r U(f;x_0,y)}{\partial x^r} = f^*_{(r)}(x_0).$$

(b) *There exist continuous functions φ and $g\in L(R)$ such that $\varphi^*_{(1)}(x_0)$ and $g^*_{(2)}(x_0)$ are finite, but the limits*

$$\lim_{(x,y)\overset{\wedge}{\longrightarrow}(x_0,0)}\frac{\partial U(\varphi;x,y)}{\partial x} \quad \textit{and} \quad \lim_{(x,y)\overset{\wedge}{\longrightarrow}(x_0,0)}\frac{\partial^2 U(g;x,y)}{\partial x^2}$$

do not exist.

Proof of Item (a). Let r be even. We have

$$\frac{\partial^r U(f;x_0,y)}{\partial x^r} = \frac{1}{\pi}\int_{-\infty}^{\infty}\frac{\partial^r P(t-x_0,y)}{\partial x^r}f(t)dt = \frac{1}{\pi}\int_{-\infty}^{\infty}\frac{\partial^r P(t,y)}{\partial x^r}f(x_0+t)dt. \quad (3.1)$$

Note that $\frac{\partial^r P(t,y)}{\partial x^r}$ is an even function of t for even r. Thus using the substitution $t=-\tau$, we obtain

$$\frac{\partial^r U(f;x_0,y)}{\partial x^r}=\frac{1}{\pi}\int\limits_{-\infty}^{\infty}\frac{\partial^r P(t,y)}{\partial x^r}f(x_0-t)dt. \tag{3.2}$$

The equalities (3.1) and (3.2) result in

$$\frac{\partial^r U(f;x_0,y)}{\partial x^r}=\frac{1}{\pi}\int\limits_{-\infty}^{\infty}\frac{\partial^r P(t,y)}{\partial x^r}\cdot\frac{f(x_0+t)+f(x_0-t)}{2}dt.$$

Owing to Statements (1) and (2), from Lemma 1.2.1 we obtain

$$\frac{\partial^r U(f;x_0,y)}{\partial x^r}=\frac{1}{\pi}\int\limits_{-\infty}^{\infty}\frac{\partial^r P(t,y)}{\partial x^r}$$

$$\times\left[\frac{\frac{f(x_0+t)+f(x_0-t)}{2}-\sum\limits_{v=0}^{\frac{r-2}{2}}\beta_{2v}\frac{t^{2v}}{(2v)!}}{\frac{t^r}{r!}}-f^*_{(r)}(x_0)\right]\frac{t^r}{r!}dt+f^*_{(r)}(x_0)$$

$$=\frac{1}{\pi}\left(\int\limits_{V_\delta}+\int\limits_{CV_\delta}\right)+f^*_{(r)}(x_0)=\frac{1}{\pi}(I_1+I_2)+f^*_{(r)}(x_0),$$

where $V_\delta=]-\delta,\delta[$. Let $\varepsilon>0$ and choose $\delta>0$ such that

$$\left|\frac{\frac{f(x_0+t)+f(x_0-t)}{2}-\sum\limits_{v=0}^{\frac{r-2}{2}}\beta_{2v}\frac{t^{2v}}{(2v)!}}{\frac{t^r}{r!}}-f^*_{(r)}(x_0)\right|<\varepsilon \tag{3.3}$$

for $|t|<\delta$.

By virtue of the above formula and Statement (3), from Lemma 1.2.1 we have the estimate $|I_1|<C\varepsilon$ for $|t|<\delta$.

Next, taking into account Statement (5), from Lemma 1.2.1 we obtain

$$|I_2|\le\frac{1}{\pi}\int\limits_{CV}\left|\frac{\partial^r P(t,y)}{\partial x^r}\right|\left[\frac{f(x_0+t)+f(x_0-t)}{2}+\sum_{v=0}^{\frac{r-2}{2}}|\beta_{2v}|\frac{|t^{2v}|}{(2v)!}\right]dt$$

$$+|f^*_{(r)}(x_0)|\int\limits_{CV}\left|\frac{\partial^r P(t,y)}{\partial x^r}\right||t^r|dt<Cy. \tag{3.4}$$

The validity of Item (a) follows from (3.3) and (3.4) (assuming ε is arbitrarily small).

Proof of Item (b). 1. We assume that $D=[-1;1]$. Let

$$\varphi(t)=\begin{cases}\sqrt{-t}, & \text{when } -1\le t\le 0,\\ \sqrt{t}, & \text{when } 0\le t\le 1,\end{cases}$$

and on the set $R|D$ we extend the function $\varphi(t)$ continuously so that $\varphi\in L(R)$. It can be easily verified that $\varphi^*_{(1)}(o)=0$. Let $(x,y)\to(o,o)$ so that $x>0$ and $y=x$. Assume that $x<\frac{1}{4}$. Then for the constructed function

$$\begin{aligned}\frac{\partial U(\varphi;x,y)}{\partial x}&=\frac{2y}{\pi}\int_0^1\frac{(t-x)\sqrt{t}}{[(t-x)^2+y^2]^2}\,dt+\frac{2y}{\pi}\int_{-1}^0\frac{(t-x)\sqrt{-t}}{[(t-x)^2+y^2]^2}\,dt+O(1)\\ &=\frac{2y}{\pi}\int_{-x}^{1-x}\frac{t\sqrt{t+x}}{(t^2+y^2)^2}\,dt-\frac{2y}{\pi}\int_0^1\frac{(t+x)\sqrt{t}}{[(t+x)^2+y^2]^2}\,dt+O(1)\\ &=\frac{2y}{\pi}\int_{-x}^{x}\frac{t\sqrt{t+x}}{(t^2+y^2)^2}\,dt+\frac{2y}{\pi}\int_x^{1-x}\frac{t(\sqrt{t+x}-\sqrt{t-x})}{(t^2+y^2)^2}\,dt+O(1)\\ &=I_1+I_2+O(1),\end{aligned}\tag{3.5}$$

where

$$I_1=\frac{2y}{\pi}\int_{-x}^{x}\frac{t\sqrt{t+x}}{(t^2+y^2)^2}\,dt=\frac{2y}{\pi}\int_0^{x}\frac{t(\sqrt{t+x}-\sqrt{x-t})}{(t^2+y^2)^2}\,dt>0,\tag{3.6}$$

$$\begin{aligned}I_2&=\frac{2x}{\pi}\int_x^{1-x}\frac{t(\sqrt{x+t}-\sqrt{t-x})}{(t^2+x^2)^2}\,dt>\frac{2x}{\pi}\int_x^{2x}\frac{t(\sqrt{x+t}-\sqrt{t-x})}{(t^2+x^2)^2}\,dt\\ &>\frac{2x}{\pi}\int_x^{2x}\frac{x(\sqrt{2x}-\sqrt{x})}{(5x)^4}\,dt=\frac{C}{\sqrt{x}}.\end{aligned}\tag{3.7}$$

It follows from (3.5) and (3.7) that

$$\frac{\partial U(\varphi;z,y)}{\partial x}\to+\infty,$$

as $(x,y)\to(o,o)$ along the chosen path.

2. We now construct the function g. Let

$$g(t)=\begin{cases}\sqrt{-t}, & \text{when } -1\le t\le 0,\\ -\sqrt{t}, & \text{when } 0\le t\le 1,\end{cases}$$

and on the set $R|D$ we extend the function $g(t)$ continuously so that $g \in L(R)$. As is easily seen, $g^*_{(2)}(o) = 0$. Let $(x, y) \to (o, o)$ so that $x > 0$, $y = x$ and $x < \frac{1}{2}$.

For this function,

$$\frac{\partial^2 U(g; x, y)}{\partial x^2} = \frac{2x}{\pi} \int_{-1}^{0} \frac{3(t-x)^2 - x^2}{[(t-x)^2 + x^2]^3} \sqrt{-t} dt - \int_{0}^{1} \frac{3(t-x)^2 - x^2}{[(t-x)^2 + x^2]^3} \sqrt{t} dt + o(1)$$

$$= \frac{2x}{\pi} \left(\int_{x}^{1+x} \frac{3t^2 - x^2}{(t^2 + x^2)^3} \sqrt{t - x} dt - \int_{-x}^{1-x} \frac{3t^2 - x^2}{(t^2 + x^2)^3} \sqrt{t + x} dt \right) + 0(1)$$

$$= \frac{2x}{\pi} \left(- \int_{x}^{1-x} \frac{3t^2 - x^2}{(t^2 + x^2)^3} (\sqrt{t+x} - \sqrt{t-x}) dt - \int_{0}^{x} \frac{3t^2 - x^2}{(t^2 + x^2)^3} \sqrt{x - t} dt \right.$$

$$\left. - \int_{0}^{x} \frac{3t^2 - x^2}{(t^2 + x^2)^3} \sqrt{x + t} dt \right) + o(1)$$

$$= \frac{2x}{\pi x^{\frac{5}{2}}} \left(- \int_{1}^{\frac{1-x}{x}} \frac{3t^2 - 1}{(1 + t^2)^3} (\sqrt{t+1} - \sqrt{t-1}) dt \right.$$

$$\left. - \int_{0}^{1} \frac{3t^2 - 1}{(1 + t^2)^3} \sqrt{1 - t} dt - \int_{0}^{1} \frac{3t^2 - 1}{(1 + t^2)^3} \sqrt{1 + t} dt \right) + o(1)$$

$$= \frac{2}{\pi x^{\frac{3}{2}}} \left(- \int_{1}^{\frac{1-x}{x}} \frac{3t^2 - 1}{(1 + t^2)^3} (\sqrt{t+1} - \sqrt{t-1}) dt \right.$$

$$\left. - \int_{1/\sqrt{3}}^{1} \frac{3t^2 - 1}{(1 + t^2)^3} (\sqrt{1+t} + \sqrt{1-t}) dt + \int_{0}^{\frac{1}{\sqrt{3}}} \frac{1 - 3t^2}{(1 + t^2)^3} (\sqrt{1+t} + \sqrt{1-t}) dt + o(1) \right)$$

$$= \frac{2}{\pi x^{3/2}} (-I_1 - I_2 + I_3) + 0(1). \tag{3.8}$$

It can be easily verified that

$$I_3 > \left(\sqrt{1 + \frac{1}{\sqrt{3}}} + \sqrt{1 - \frac{1}{\sqrt{3}}} \right) \int_{0}^{\frac{1}{\sqrt{3}}} \frac{1 - 3t^2}{(1 + t^2)} dt$$

$$= \frac{9}{16\sqrt{3}} \left(\sqrt{1 + \frac{1}{\sqrt{3}}} + \sqrt{1 - \frac{1}{\sqrt{3}}} \right). \tag{3.9}$$

$$I_1 < \sqrt{2}\int\limits_1^\infty \frac{3t^2-1}{(1+t^2)^3}\,dt = \frac{\sqrt{2}}{4}. \tag{3.10}$$

$$I_2 < \left(\sqrt{1+\frac{1}{\sqrt{3}}}+\sqrt{1-\frac{1}{\sqrt{3}}}\right)\int\limits_{1/\sqrt{3}}^1 \frac{3t^2-1}{(1+t^2)^3}\,dt$$

$$= \frac{9-4\sqrt{3}}{16\sqrt{3}}\left(\sqrt{1+\frac{1}{\sqrt{3}}}+\sqrt{1-\frac{1}{\sqrt{3}}}\right). \tag{3.11}$$

The expressions (3.9), (3.10) and (3.11) yield

$$I_3 - I_1 - I_2 > \frac{1}{4}\left(\sqrt{1+\frac{1}{\sqrt{3}}}+\sqrt{1-\frac{1}{\sqrt{3}}}-\sqrt{2}\right) > 0. \tag{3.12}$$

It follows from (3.8) and (3.12) that

$$\frac{\partial^2 U(g;x,y)}{\partial x^2} > \frac{1}{2\pi x^{3/2}}\left(\sqrt{1+\frac{1}{\sqrt{3}}}+\sqrt{1-\frac{1}{\sqrt{3}}}-\sqrt{2}\right),$$

whence

$$\frac{\partial^2 U(g;x,y)}{\partial x^2} \to +\infty$$

as $(x,y)\to(o,o)$ along the chosen path.

The theorem is proved. □

Let $F(x)$ be an undefined integral of a function $f\in L(R)$, i.e.,

$$F(x) = \int\limits_{-\infty}^x f(t)dt.$$

Corollary 1.3.1. *At every point x_0 at which there exists a finite $F^*_{(1)}(x_0)$ (see (1.1)), we have*

$$\lim_{y\to 0+} U(f;x_0,y) = F^*_{(1)}(x_0).$$

Proof. Integrating by parts, we obtain

$$U(f;x,y) = \frac{1}{\pi}\int\limits_{-\infty}^\infty P(t-x,y)f(t)dt$$

$$= -\frac{1}{\pi}\int\limits_{-\infty}^\infty F(t)\frac{\partial P(t-x,y)}{dt}\,dt = \frac{\partial}{\partial x}U(F;x,y). \tag{3.13}$$

Hence by virtue of Theorem 1.3.1, it follows that Corollary 1.3.1 is valid. □

Theorem 1.3.2. (a) *If at the point x_0 there exists a finite $f_{(r)}(x_0)$, then*

$$\lim_{(x,y)\overset{\wedge}{\longrightarrow}(x_0,0)} \frac{\partial^r U(f;x,y)}{\partial x^r} = f_{(r)}(x_0).$$

(b) *There exists a function $g \in L(R)$ such that $g'(x_0)$ is finite, but the limit*

$$\lim_{(x,y)\to(x_0,0)} \frac{\partial U(g;x,y)}{\partial x}$$

does not exist.

Proof of Item (a). Let $x_0 = 0$. By Statements (1) and (2) of Lemma 1.2.1, we have

$$\frac{\partial^r U(f;x,y)}{\partial x^r} = \frac{1}{\pi}\int_{-\infty}^{\infty} \frac{\partial^r P(t-x,y)}{\partial x^r} f(t)dt$$

$$= \frac{1}{\pi}\int_{-\infty}^{\infty} \frac{\partial^r P(t-x,y)}{\partial x^r}\left[\frac{f(t)-\sum\limits_{v=0}^{r-1}\alpha_v\frac{t^v}{v!}}{t^r/r!} - f_{(r)}(o)\right]\frac{t^r}{r!}dt + f_{(r)}(o)$$

$$= \frac{1}{\pi}\left(\int_{V_\delta} + \int_{CV_\delta}\right) + f_{(r)}(o) = \frac{1}{\pi}(I_1+I_2) + f_{(r)}(o), \tag{3.14}$$

where $v_\delta =]-\delta,\delta[$. Let $\varepsilon > 0$. We choose $\delta > 0$ such that

$$\left[\frac{f(t)-\sum\limits_{v=0}^{r-1}\alpha_v\frac{t^v}{v!}}{t^r/r!} - f_{(r)}(o)\right] < \varepsilon, \quad \text{for} \quad |t| < \delta.$$

Hence by Statement (4) of Lemma 1.2.1, we have

$$|I_1| < C\varepsilon, \quad for \quad |t| < \delta \quad \text{and} \quad \frac{y}{|x|} \geq C > 0. \tag{3.15}$$

Furthermore, in view of Statement (5) of Lemma 1.2.1, we have

$$|I_2| \leq \frac{1}{\pi}\int_{CV}\left|\frac{\partial^r P(t-x,y)}{\partial x^r}\right|\left[|f(t)| + \sum_{v=0}^{r-1}|\alpha_v|\frac{|t^v|}{v!}\right]dt$$
$$+|f_{(r)}(o)|\int_{CV}\left|\frac{\partial^r P(t-x,y)}{\partial x^r}\right||t^r|dt < Cy, \tag{3.16}$$

for $|x| < \frac{\delta}{2}$ and $\frac{y}{|x|} \geq C > 0$.

By virtue of (3.15), (3.16), from (3.14) we obtain

$$\lim_{(x,y)\overset{\wedge}{\longrightarrow}(0,0)} \frac{\partial^r U(f;x,y)}{\partial x^r} = f_{(r)}(o).$$

Proof of Item (b). Let

$$g(t) = \begin{cases} \sqrt{|t^3|}\sin\dfrac{1}{t}, & \text{when} \quad t \in [-1;0[\cup]0;1], \\ 0, \quad \text{when} \quad t=0 & \text{and} \quad t \in]-\infty,-1[\cup]1,+\infty[. \end{cases}$$

Clearly, $g'(o) = 0$, but as for $\overline{g}'(o) = \lim\limits_{(t,x)\to(0,0)} \frac{f(x+t)-f(x)}{t}$, it does not exist. Indeed, let $t = x^2$ and $0 < x < \frac{1}{2}$. Then

$$\begin{aligned} &\lim_{(x,t)\to(0,0)} \frac{\sqrt{|x+t|^3}\sin\frac{1}{x+t} - \sqrt{|x^3|}\sin\frac{1}{x}}{t} \\ &= \lim_{x\to0} \frac{\sqrt{(x+x^2)^3}\sin\frac{1}{x+x^2} - \sqrt{x^3}\sin\frac{1}{x}}{x^2} \\ &= \lim_{x\to0} \frac{1}{\sqrt{x}}\Big[\sqrt{(1+x)^3}\sin\frac{1}{x+x^2} - \sin\frac{1}{x}\Big]. \end{aligned}$$

If $x_n = \frac{1}{2n\pi}$, then $\frac{1}{x_n+x_n^2} = 2n\pi - \frac{2n\pi}{2n\pi+1}$, and hence

$$\begin{aligned} &\lim_{x\to0} \frac{1}{\sqrt{x}}\Big[\sqrt{(1+x)^3}\sin\frac{1}{x+x^2} - \sin\frac{1}{x}\Big] \\ &= \lim_{n\to\infty} \sqrt{2n\pi}\sqrt{(1+\frac{1}{2n\pi})^3\sin\frac{2n\pi}{2n\pi+1}} = +\infty. \end{aligned}$$

Thus $\overline{g}'(o)$ does not exist.

Let $(x,y) \to (0,0)$ so that $x > 0$, $y = x^2$ and $x < \frac{1}{2}$. Then for the above-constructed function

$$U(g;x,y) = \frac{2y}{\pi}\int_{-\infty}^{\infty} \frac{(t-x)g(t)dt}{[(t-x)^2+y^2]^2} = \frac{2y}{\pi}\int_{-1}^{1} \frac{(t-x)\sqrt{|t^3|}\sin\frac{1}{t}dt}{[(t-x)^2+y^2]^2}$$

$$= \frac{2y}{\pi}\Big\{\int_0^{1-x} \frac{\sqrt{(x+t)^3}\sin\frac{1}{x+t}}{(t^2+y^2)^2}\,dt - \int_0^{1+x} \frac{t\sqrt{|x-t|^3}\sin\frac{1}{x-t}}{(t^2+y^2)^2}\,dt\Big\}$$

$$= \frac{2y}{\pi}\int_0^{1-x} \frac{t[\sqrt{(x+t)^3}\sin\frac{1}{x+t} - \sqrt{|x-t|^3}\sin\frac{1}{x-t}]}{(t^2+y^2)^2}\,dt + o(1)$$

$$= \frac{2x^2}{\pi}\Big\{\int_0^x + \int_x^{1-x}\Big\} = \frac{2}{\pi}(I + I_1).$$

$$I_1 = x^2 \int_x^{1-x} \frac{t[\sqrt{(x+t)^3}\sin\frac{1}{x+t} - \sqrt{(t-x)^3}\sin\frac{1}{x-t}]}{(t^2+x^4)^2}\,dt$$

$$= [t = x\tau] = \sqrt{x^3}\int_1^{\frac{1-x}{x}} \frac{t[\sqrt{(1+t)^3}\sin\frac{1}{x+xt} - \sqrt{(t-1)^3}\sin\frac{1}{x-xt}]}{(t^2+x^2)^2}\,dt.$$

whence it follows that

$$|I_1| < C\sqrt{x^3}\int_1^\infty \frac{t\sqrt{(1+t)^3}}{t^4}\,dt < C\sqrt{x^3}\int_1^\infty \frac{t^{\frac{5}{2}}}{t^4}\,dt = C\sqrt{x^3}\int_1^\infty \frac{dt}{t^{3/2}}.$$

This implies that

$$\lim_{x\to 0} I_1 = 0.$$

To investigate the integral

$$I(x) = x^2 \int_0^x \frac{t[\sqrt{(x+t)^3}\sin\frac{1}{x+t} - \sqrt{(x-t)^3}\sin\frac{1}{x-t}]}{(t^2+x^4)^2}\,dt$$

$$x^2 \int_0^x \frac{t[\sqrt{(x+t)^3}\sin\frac{1}{x+t} + \sqrt{(x-t)^3}\sin\frac{1}{t-x}]}{(t^2+x^4)^2}\,dt \tag{3.17}$$

we perform the Maclaurin-series expansion of the functions

$$\sin\frac{1}{t+x} \quad \text{and} \quad \sin\frac{1}{t-x}.$$

We have

$$\sin\frac{1}{t+x} = \sin\frac{1}{x} + \Big(-\frac{1}{x^2}\cos\frac{1}{x}\Big)\cdot\frac{t}{1!} + \Big(\frac{2}{x^3}\cos\frac{1}{x} - \frac{1}{x^4}\sin\frac{1}{x}\Big)\cdot\frac{t^2}{2!}$$
$$+\Big[\Big(-\frac{6}{x^4}+\frac{1}{x^6}\Big)\cos\frac{1}{x} + \frac{6}{x^5}\sin\frac{1}{x}\Big]\cdot\frac{t^3}{3!}$$
$$+\Big[\Big(\frac{24}{x^5}-\frac{12}{x^7}\Big)\cos\frac{1}{x} + \Big(-\frac{36}{x^6}+\frac{1}{x^8}\Big)\sin\frac{1}{x}\Big]\cdot\frac{t^4}{4!}$$
$$+\Big[\Big(-\frac{120}{x^6}+\frac{120}{x^8}-\frac{1}{x^{10}}\Big)\cos\frac{1}{x} + \Big(\frac{240}{x^7}-\frac{20}{x^9}\Big)\sin\frac{1}{x}\Big]\cdot\frac{t^5}{5!}+\cdots$$

and

$$\sin\frac{1}{t-x} = -\sin\frac{1}{x} + \Big(-\frac{1}{x^2}\cos\frac{1}{x}\Big)\cdot\frac{t}{1!} - \Big(\frac{2}{x^3}\cos\frac{1}{x} - \frac{1}{x^4}\sin\frac{1}{x}\Big)\cdot\frac{t^2}{2!}$$

$$+\Big[\Big(-\frac{6}{x^4}+\frac{1}{x^6}\Big)\cos\frac{1}{x}+\frac{6}{x^5}\sin\frac{1}{x}\Big]\cdot\frac{t^3}{3!}$$
$$-\Big[\Big(\frac{24}{x^5}-\frac{12}{x^7}\Big)\cos\frac{1}{x}+\Big(-\frac{36}{x^6}+\frac{1}{x^8}\Big)\sin\frac{1}{x}\Big]\cdot\frac{t^4}{4!}$$
$$+\Big[\Big(-\frac{120}{x^6}+\frac{120}{x^8}-\frac{1}{x^{10}}\Big)\cos\frac{1}{x}+\Big(\frac{240}{x^7}-\frac{20}{x^9}\Big)\sin\frac{1}{x}\Big]\cdot\frac{t^5}{5!}-\cdots$$

Substituting these expansions into (3.17), we can see that $\lim\limits_{x\to 0+} I(x)$ does not exist, and hence

$$\lim\frac{\partial U(g;x,y)}{\partial x}$$

does not exist as $(x,y)\to(o,o)$ along the chosen path.

The theorem is proved. □

Corollary 1.3.2. *Let $F(x)=\int_{-\infty}^{x} f(t)dt$. At every point x_0 at which $F'(x_0)=f(x_0)$ exists and is finite (i.e., at every point x_0, where $f(x_0)$ is a finite derivative of its undefined integral. Hence the Poisson integral $U(f;x,y)$ of the function f tends almost everywhere) to $f(x_0)$ as $(x,y)\to(x_0,0)$ along the non-tangential path.*

The validity of this statement follows from Theorem 1.3.2 and equality (3.13).

Theorem 1.3.3. *If at the point x_0 there exists a finite $\overline{f}^{*}_{(r)}(x_0)$, then*

$$\lim_{(x,y)\to(x_0,o)}\frac{\partial^r U(f;x,y)}{\partial x^r}=\overline{f}^{*}_{(r)}(x_0).$$

Proof. Let r be an even number. Then by Statements (1) and (2) of Lemma 1.2.1, we have

$$\frac{\partial^r U(f;x,y)}{\partial x^r}=\frac{1}{\pi}\int_{-\infty}^{\infty}\frac{\partial^r P(t,y)}{\partial x^r}$$
$$\times\left[\frac{\frac{f(x+t)+f(x-t)}{2}-\sum\limits_{v=0}^{\frac{r-2}{2}}\beta_{2v}(x)\frac{t^{2v}}{(2v)!}}{\frac{t^r}{r!}}-\overline{f}^{*}_{(r)}(x_0)\right]\frac{t^r}{r!}\,dt+\overline{f}^{*}_{(r)}(x_0).$$

This equality, on the basis of Lemma 1.2.1, shows that the statement of the theorem is valid. □

Corollary 1.3.3. *If at the point x_0 there exists a finite $\overline{f}_{(r)}(x_0)$, then*

$$\lim_{(x,y)\to(x_0,o)}\frac{\partial^r U(f;x,y)}{\partial x^r}=\overline{f}_{(r)}(x_0).$$

Lemma 1.3.1. *Let* $F(x) = \int\limits_{-\infty}^{x} f(t)dt$*. If* $f(x)$ *is continuous at the point* x_0*, then* $\overline{F}_{(1)}(x_0) = f(x_0)$ *(see (1.3)).*

Indeed,

$$\frac{F(x+h) - F(x)}{h} = \frac{1}{h} \int\limits_{x}^{x+h} f(t)dt = \frac{1}{h} \int\limits_{x}^{x+h} [f(t) - f(x_0)]dt + f(x_0).$$

This implies that Lemma 1.3.1 is valid.

From Lemma 1.3.1 and Corollary 1.3.3 we have

Corollary 1.3.4. *At every point, at which* $\overline{F}_{(1)}(x_0) = f(x_0)$ *exists and is finite, (hence everywhere for a continuous function) we have*

$$\lim_{(x,y)\to(x_0,0)} U(f;x,y) = f(x_0).$$

Remark. The above theorems are valid for the generalized C_1-derivatives defined by the equalities

1. $C_1 f_{(r)}(x_0) = \lim\limits_{h\to 0+} \dfrac{(r+1)!}{h^{r+1}} \int\limits_0^h \left[f(x_0+t) - \sum\limits_{v=0}^{r-1} \alpha_v \dfrac{t^v}{v!} \right] dt;$
2. $C_1 \overline{f}_{(r)}(x_0) = \lim\limits_{(x,h)\to(x_0,0)} \dfrac{(r+1)!}{h^{r+1}} \int\limits_0^h \left[f(x+t) - \sum\limits_{v=0}^{r-1} \alpha_v(x) \dfrac{t^v}{v!} \right] dt;$ where $\alpha_v(x)$ and $v = \overline{0, r-1}$ are defined in the neighborhood of the point x_0, and there exist limits $\lim\limits_{x\to x_0} \alpha_v(x) = \alpha_v$.
3. Symmetric Derivatives. For odd r's,

$$C_1 f^*_{(r)}(x_0) = \lim_{h\to 0} \frac{(r+1)!}{h^{r+1}} \int\limits_0^h \left[\frac{f(x_0+t) - f(x_0-t)}{2} - \sum_{v=1}^{\frac{r-2}{2}} \beta_{2v-1} \frac{t^{2v-1}}{(2v-1)!} \right] dt.$$

For even r's,

$$C_1 f^*_{(r)}(x_0) = \lim_{h\to 0} \frac{(r+1)!}{h^{r+1}} \int\limits_0^h \left[\frac{f(x_0+t) + f(x_0-t)}{2} - \sum_{v=0}^{\frac{r-2}{2}} \beta_{2\nu} \frac{t^{2v}}{(2v)!} \right] dt.$$

4. Symmetric Strong Derivatives. For odd r's,

$$C_1 \overline{f}^*_{(r)}(x_0) = \lim_{(x,h)\to(x_0,0)} \frac{(r+1)!}{h^{r+1}} \int\limits_0^h \left[\frac{f(x+t) - f(x-t)}{2} - \sum_{v=1}^{\frac{r-1}{2}} \beta_{2t-1}(x) \frac{t^{2\nu-1}}{(2v-1)!} \right] dt,$$

for even r's,

$$C_1\overline{f}^*_{(r)}(x_0) = \lim_{(x,h)\to(x_0,0)} \frac{(r+1)!}{h^{r+1}} \int_0^h \left[\frac{f(x+t)+f(x-t)}{2} - \sum_{v=0}^{\frac{r-2}{2}} \beta_{2v}(x)\frac{t^{2v}}{(2v)!}\right] dt,$$

$\beta_v(x)$ are defined in the neighborhood of the point x_0, and there exist limits $\lim\limits_{x\to x_0} \beta_v(x) = \beta_v$.

1.4 The Dirichlet Problem for a Half-Plane

The Dirichlet problem for a half-plane R^2_+ is formulated as follows: Given a function $f(t)$ on R, find the function $U(x,y)$, harmonic (i.e., satisfying the Laplace equation $\dfrac{\partial^2 U}{\partial x^2} + \dfrac{\partial^2 U}{\partial y^2} = 0$) in the domain R^2_+ whose values tend to those of $f(\overline{x})$ when the point $M(x,y) \in R^2$ tends arbitrarily to the points $P(\overline{x}, o) \in R$.

It can be easily verified that the Poisson integral $U(f;x,y)$ is harmonic in the domain R^2_+.

Based on the results obtained in Section 1.3, we can conclude that:

(1) If $f(t) \in \widetilde{L}(R)$ is a continuous function, then the Poisson integral $U(f;x,y)$ is a solution of the Dirichlet problem in the sense that for all $\overline{x} \in R$,

$$U(f;x,y) \to f(\overline{x})$$

no matter how the point $M(x,y)$ tends to $P(\overline{x}, o)$ provided only that it remains in R^2_+.

This follows from Corollary 1.3.4.

(2) If $f(t) \in \widetilde{L}(R)$, then the Poisson integral $U(f;x,y)$ is a solution of the Dirichlet problem in the sense that almost for all $\overline{x} \in R$,

$$U(f;x,y) \to f(\overline{x})$$

when the point $M(x,y) \in R^2_+$ tends to the point $P(\overline{x}, 0)$ along a non-tangential path.

This follows from Corollary 1.3.2.

In this section, the Dirichlet problem is solved for the case where the boundary function is measurable and finite almost everywhere on R, i.e., in N. N. Luzin's formulation ([43], p. 86).

Theorem 1.4.1. *Let f be an arbitrary measurable and finite function almost everywhere on R. Then there exists a harmonic function $U(x,y)$ in R^2_+ such that*

$$\lim_{(x,y)\overset{\wedge}{\longrightarrow}(\overline{x},0)} U(x,y) = f(\overline{x})$$

almost everywhere on R.

Proof. Using Luzin's theorem on finding primitive functions (see [43], p. 78), we construct, for f in R, a continuous bounded function F such that the inequality

$$F'(x) = f(x)$$

is fulfilled almost everywhere on R.

Consider the function

$$U(x, y) = \frac{y}{\pi} \int_R F(t) \frac{\partial}{\partial x} \left[\frac{1}{(t-x)^2 + y^2} \right] dt.$$

It is not difficult to verify that the function $U(x, y)$ is harmonic in R^2_+. By Theorem 1.3.2, if at the point $\overline{x} \in R$ there exists $F'(\overline{x})$, then

$$\lim_{(x,y) \overset{\wedge}{\longrightarrow} (\overline{x},0)} U(x, y) = F'(\overline{x}).$$

Since $F'(x) = f(x)$ almost everywhere on R, Theorem 1.4.1 is proved. □

Chapter 2

Boundary Properties of Derivatives of the Poisson Integral for a Circle

2.1 Notation and Definitions

Let 2π be a periodic function $f \in L(-\pi;\pi)$, and

$$\frac{1}{2}a_0 + \sum_{v=1}^{\infty}(a_v \cos vx + b_v \sin vx) \tag{1.1}$$

its Fourier series, i.e.,

$$a_v = \frac{1}{\pi}\int_{-\pi}^{\pi} f(t)\cos vt dt, \quad b_v = \frac{1}{\pi}\int_{-\pi}^{\pi} f(t)\sin vt dt \tag{1.2}$$

$$(v = 0, 1, 2, \dots).$$

We denote the series (1.1) by the symbol $S[f]$.

The abelian means for $S[f]$, denoted by $U(f;r,x)$, are defined by the equality

$$U(f;r,x) = \frac{1}{2}a_0 + \sum_{v=1}^{\infty}(a_v \cos vx + b_v \sin vx)r^v, \quad (0 < r < 1).$$

Taking (1.2) into account, we can easily show that

$$U(f;r,x) = \frac{1}{\pi}\int_{-\pi}^{\pi} f(t)P(r,t-x)dt = \frac{1}{\pi}\int_{-\pi}^{\pi} f(x+t)P(r,t)dt, \tag{1.3}$$

where

$$P(r,t)=\frac{1}{2}+\sum_{v=1}^{\infty} r^v \cos vt=\frac{1}{2}\cdot\frac{1-r^2}{1-2r\cos t+r^2}$$
$$=\frac{1}{2}\cdot\frac{1-r^2}{(1-r)^2+4r\sin^2\frac{t}{2}}.$$

The right-hand side of the equality (1.3) is usually called the Poisson integral of f for a circle. Hence the expressions "abelian means of the series $S[f]$" and "the Poisson integral of the function f for a circle" are synonyms.

The symbol $(r,x)\xrightarrow{\wedge}(1,x_0)$, $(re^{ix}\xrightarrow{\wedge}e^{ix_0})$ denotes ([2], p.30) that the point $M(r,x)$ tends to $P(1,x_0)$ along a tangential path, i.e., there exists a positive number C, such that $\frac{\rho}{1-r}<C$, where ρ is the distance between the points $M(r,x)$ and $P(1,x_0)$.

Under $M(r,x)\to P(1,x_0)$ we mean that the point $M(r,x)$ tends to $P(1,x_0)$ arbitrarily, remaining in the unit circumference.

2.2 Auxiliary Statements

It is easy to verify that

$$\frac{\partial P(r,t)}{\partial t}=-r(1-r^2)\frac{A_1(r,t)}{[(1-r)^2+4r\sin^2\frac{t}{2}]^2},$$

where

$$A_1(r,t)=\sin t\frac{\partial^2 P(r,t)}{\partial t^2}=-r(1-r^2)\frac{A_2(r,t)}{[(1-r)^2+4r\sin^2\frac{t}{2}]^3},$$

where $A_2(r,t)=(1-r)^2\cos t+4r\cos t\sin^2\frac{t}{2}-4r\sin^2 t$.

$$\frac{\partial^3 P(r,t)}{\partial t^3}=-r(1-r^2)\frac{A_3(r,t)}{[(1-r)^2+4r\sin^2\frac{t}{2}]^4},$$

where

$$A_3(r,t)=-(1-r)^4\sin t-8r(1-r)^2\sin t\sin^2\frac{t}{2}$$
$$-6r(1-r)^2\sin 2t-16r^2\sin t\sin^4\frac{t}{2}-24r^2\sin 2t\sin^2\frac{t}{2}+24r^2\sin^3 t.$$

. . . .

$$\frac{\partial^n P(r,t)}{\partial t^n}=-r(1-r^2)\frac{A_n(r,t)}{[(1-r)^2+4r\sin^2\frac{t}{2}]^{n+1}}.$$

The following statements are valid:

1) $A_n(r,-t) = (-1)^n A_n(r,t)$. (2.1)
2) All terms in $A_n(r,t)$ with respect to $(1-r)$ and $\sin t$ are of degree $\geq n$.
The validity of Statement 1) is easily verified.
Statement 2) is valid for $n = 1, 2$ and 3. Assume that this is the case for $A_n(r,t)$. Then

$$A_{n+1}(r,t) = A_n'(r,t)\left[(1-r)^2 + 4r\sin^2\frac{t}{2}\right] - 2(n+1)r\sin t \cdot A_n(r,t).$$

This equality shows that Statement 2) is valid.
From Statement 2) we have

$$|A_n(r,t)| \leq I_n(r,t), \tag{2.2}$$

where $I_n(r,t)$ is a homogeneous polynomial of degree n of $(1-r,t)$ with a positive coefficient.

Lemma 2.2.1. *For any $n \in N$, the following statements are valid:*
1) $\int\limits_{-\pi}^{\pi} \left|\frac{\partial^n P(r,t)}{\partial t^n}\right| |t|^n dt = O(1)$,
2) $\int\limits_{-\pi}^{\pi} \left|\frac{\partial^n P(r,t-x)}{\partial t^n}\right| |t|^n dt = O(1)$ *for* $\frac{|x|}{1-r} \leq C$,
3) $\lim\limits_{r\to 1} \max\limits_{0<\delta\leq t\leq\pi} \left|\frac{\partial^n P(r,t)}{\partial t^n}\right| = 0$.

Proof. Taking into account (2.2) and the inequality $|\sin t| \geq \frac{2}{\pi}|t|$ for $|t| \leq \frac{\pi}{2}$, we obtain

$$\int\limits_{-\pi}^{\pi} \left|\frac{\partial^n P(r,t)}{\partial t^n}\right| |t|^n dt \leq 2(1-r)\int\limits_{-\pi}^{\pi} \frac{|A_n(r,t)||t|^n}{[(1-r)^2 + 4r\sin^2\frac{t}{2}]^{n+1}}\, dt$$

$$\leq 2(1-r)\pi^{2(n+1)} \int\limits_{-\pi}^{\pi} \frac{I_n(r,t)|t|^n dt}{[(1-r)^2\pi^2 + 4rt^2]^{n+1}}. \tag{2.3}$$

After substituting $t = (1-r)\tau$, (2.3) yields

$$\int\limits_{-\pi}^{\pi} \left|\frac{\partial^n P(r,t)}{\partial t^n}\right| |t|^n dt < C(1-r)\int\limits_{-\frac{\pi}{1-r}}^{\frac{\pi}{1-r}} \frac{(1-r)^{2n+1}\left(\sum_{v=0}^{n} C_v |t|^v\right)|t|^n dt}{(1-r)^{2n+2}(1+t^2)^{n+1}}$$

$$< C\int\limits_{0}^{\infty} \frac{\sum_{v=0}^{n} C_v t^{v+n}}{(1+t^2)^{n+1}}\, dt = O(1).$$

Statement 2) follows from 1) if we take into account the conditions $\frac{|x|}{1-r} < C$ (non-tangential approach of the point (r,x) to the point $(1,0)$) and the inequality

$$|a+b|^p \leq 2^p(|a|^p + |b|^p), \quad p \geq 1$$

(for the proof of Statement 4), see Lemma 1.2.1). □

Lemma 2.2.2. *If from the conditions $g(o) = 0$ and $g^{(k)}(o) = 0$, $k = 1,2,\ldots,n$ follows*

$$\lim_{(r,x)\to(1,0)} \frac{\partial^n U(g;r,x)}{\partial x^n} = 0$$

then the condition $f^{(n)}(o) = S$ results in

$$\lim_{(r,x)\to(1,0)} \frac{\partial^n U(f;r,x)}{\partial x^n} = S.$$

Proof. First of all, we have to show that if $\alpha_0, \alpha_1, \alpha_2, \ldots, \alpha_n$ are the given numbers, then there exists a trigonometric polynomial $T_n(t) = \sum_{v=0}^n a_v e^{ivt}$ ([69], p. 287; [67], p.2 25), such that

$$T_n(o) = \alpha_0 \quad \text{and} \quad T_n^{(k)}(o) = \alpha_k, \quad k = 1,2,\ldots,n, \quad ??$$

$$\begin{cases} a_0 + a_1 + a_2 + \cdots + a_n = \alpha_0 \\ i(a_1 + 2a_2 + \cdots + na_n) = \alpha_1 \\ i^2(a_1 + 2^2a_2 + \cdots + n^2a_n) = \alpha_2 \\ \cdots\cdots\cdots\cdots\cdots\cdots \\ i^n(a_1 + 2^na_2 + \cdots + n^na_n) = \alpha_n. \end{cases}$$

The above system has a unique solution since its (Vandermonde) determinant is not equal to zero.

Let now $g(t) = f(t) - T_n(t)$, where $T_n(o) = f(o)$ and $T_n^{(k)}(o) = f^{(k)}(o)$, $k = 1,2,\ldots,n$ $(T_n^{(n)}(o) = S)$.

Moreover,

$$U(g;r,x) = U(f;r,x) - U(T_n;r,x) = U(f;r,x) - \sum_{v=0}^n a_v r^v e^{ivx}$$

whence

$$\frac{\partial^n U(g;r,x)}{\partial x^n} = \frac{\partial^n U(f;r,x)}{\partial x^n} - \frac{\partial^n}{\partial x^n}\left(\sum_{v=0}^n a_v r^v e^{ivx}\right).$$

Hence we obtain

$$\lim_{(r,x)\to(1,0)} \frac{\partial^n U(g;r,x)}{\partial x^n} = \lim_{(r,x)\to(1,0)} \frac{\partial^n U(f;r,x)}{\partial x^n} - S = 0.$$

Lemma 2.2.2 is proved. □

2.3 Boundary Properties of Derivatives of the Poisson Integral for a Circle

The boundary properties of derivatives of the Poisson integral were studied by P. Fatou ([33]). In particular, he proved the following

Theorem A. *If there exists $f'(x_0)$ and is finite, then*

$$\lim_{re^ix \xrightarrow{\wedge} e^{ix_0}} \frac{\partial U(f;r,x)}{\partial x} = f'(x_0)$$

([2], p. 156; [34], p. 167).

Theorem B. *If $f^*_{(1)}(x_0)$ exists and is finite or infinite (see Ch.* I, (1.1)), *then*

$$\lim_{r\to 1-} \frac{\partial U(f;r,x_0)}{\partial x} = f^*_{(1)}(x_0),$$

where $f^*_{(1)}(x_0)$ is the first symmetric derivative of $f(x)$ at the point x_0 ([2], p. 161; [34], p. 166).

In [89], the author constructed a continuous 2π-periodic function $f(x)$ such that $f^*_{(1)}(x_0) = 0$, but the limit

$$\lim_{re^ix \xrightarrow{\wedge} e^{ix_0}} \frac{\partial U(f;r,x)}{\partial x}$$

does not exist. Thus it is shown that Theorem B cannot be strengthened in a sense of the existence of an angular limit.

In this section, we prove analogues of the Fatou theorem for generalized derivatives of any order ([98]) and show that they cannot be strengthened in a certan sense.

Theorem 2.3.1. (a) *If at the point x_0 there exists a finite function $f^*_{(n)}(x_0)$, then*

$$\lim_{r\to 1-} \frac{\partial^n U(f;r,x_0)}{\partial x^n} = f^*_{(n)}(x_0).$$

(b) *There exist 2π-periodic continuous functions $\varphi(t)$ and $g(t)$ such that $\varphi^*_{(1)}(x_0)$ and $g^*_{(2)}(x_0)$ are finite, but the limits*

$$\lim_{(r,x)\xrightarrow{\wedge}(1,x_0)} \frac{\partial U(\varphi;r,x)}{\partial x}, \qquad \lim_{(r,x)\xrightarrow{\wedge}(1,x_0)} \frac{\partial^2 U(g;r,x)}{\partial x^2}$$

do not exist.

Proof of Item (a). Let n be an even number. We have

$$\frac{\partial^n U(f;r,x_0)}{\partial x^n}=\frac{1}{\pi}\int\limits_{-\pi}^{\pi}\frac{\partial^n P(r;t-x_0)}{\partial x^n}f(t)dt=\frac{1}{\pi}\int\limits_{-\pi}^{\pi}\frac{\partial^n P(r,t)}{\partial x^n}f(x_0+t)dt. \qquad (3.1)$$

Taking into account (2.1), from (3.1) we find that

$$\frac{\partial^n U(f;r,x_0)}{\partial x^n}=\frac{1}{\pi}\int\limits_{-\pi}^{\pi}\frac{\partial^n P(r,t)}{\partial x^n}\cdot\frac{f(x_0+t)+f(x_0-t)}{2}dt. \qquad (3.2)$$

By Lemma 2.2.2, it can be assumed that $f(x_0)=f^*_{(1)}(x_0)=f^*_{(2)}(x_0)=\cdots=f^*_{(n)}(x_0)=0$. Let $\varepsilon>0$ and choose $\delta>0$ such that

$$\frac{f(x_0+t)+f(x_0-t)}{2}<\varepsilon|t^n| \quad \text{for} \quad |t|<\delta. \qquad (3.3)$$

By Statements 1) and 3) of Lemma 2.2.1 and inequality (3.3), from (3.2) we obtain

$$\Big|\frac{\partial^n U(f;r,x_0)}{\partial x^n}\Big|\leq\frac{\varepsilon}{\pi}\int\limits_{-\delta}^{\delta}\Big|\frac{\partial^n P(r,t)}{\partial t^n}\Big||t^n|dt$$

$$+C\max_{\delta\leq 1\leq\pi}\Big|\frac{\partial^n P(r,t)}{\partial t^n}\Big|\int\limits_{-\pi}^{\pi}|f(t)|dt=o(1) \quad \text{as} \quad r\to 1-.$$

Thus Item (a) is proved.

Proof of Item (b). Define the function $\varphi(t)$ as follows:

$$\varphi(t)=\begin{cases}\sqrt{-t}, & \text{when} \quad -\pi<t<0,\\ \sqrt{t}, & \text{when} \quad 0\leq t<\pi,\\ 0, & \text{when} \quad t=\pm\pi\end{cases}$$

and $\varphi(2k\pi+t)=\varphi(t)$, $k=\pm1,\pm2,\dots$. It is not difficult to verify that $\varphi^*_{(1)}(o)=0$. For this function we have

$$\frac{\partial U(\varphi;r,x)}{\partial x}=\frac{r(1-r^2)}{\pi}\int\limits_{-\pi}^{\pi}\frac{\sin(t-x)\varphi(t)dt}{[(1-r)^2+4r\sin^2\frac{t-x}{2}]^2}$$

$$=C(1-r)\Big\{\int\limits_{-\pi}^{0}\frac{\sqrt{-t}\sin(t-x)dt}{[(1-r)^2+4r\sin^2\frac{t-x}{2}]^2}+\int\limits_{0}^{\pi}\frac{\sqrt{t}\sin(t-x)dt}{[(1-r)^2+4r\sin^2\frac{t-x}{2}]^2}\Big\}$$

$$= C(1-r)\Big\{ -\int\limits_{x}^{\pi+x} \frac{\sqrt{t-x}\sin t dt}{[(1-r)^2+4r\sin^2\frac{t}{2}]^2} + \int\limits_{-x}^{\pi-x} \frac{\sqrt{t+x}\sin t dt}{[(1-r)^2+4r\sin^2\frac{t}{2}]^2}\Big\}$$

$$= C(1-r)\Big\{ \int\limits_{-x}^{x} \frac{\sqrt{t+x}\sin t dt}{[(1-r)^2+4r\sin^2\frac{t}{2}]^2}$$

$$+ \int\limits_{x}^{\pi-x} \frac{(\sqrt{t+x}-\sqrt{t-x})\sin t dt}{[(1-r)^2+4r\sin^2\frac{t}{2}]^2} - \int\limits_{\pi-x}^{\pi+x} \frac{\sqrt{t-x}\sin t dt}{[(1-r)^2+4r\sin^2\frac{t}{2}]^2}\Big\}$$

$$= C(1-r)(I_1+I_2+I_3). \tag{3.4}$$

Let $x_0 = 0$ and $(r,x) \to (1,0)$ so that $0 \le x < \frac{\pi}{4}$ and $\sin x = 1-r$. It can easily show that

$$I_1 > 0. \tag{3.5}$$

Next,

$$I_3 = -\int\limits_{\pi-x}^{\pi+x} \frac{\sqrt{t-x}\sin t dt}{[(1-r)^2+4r\sin^2\frac{t}{2}]^2} = \int\limits_{-x}^{x} \frac{\sqrt{\pi+t-x}\sin t dt}{[(1-r)^2+4r\cos^2\frac{t}{2}]^2}$$

$$= \int\limits_{0}^{x} \frac{(\sqrt{\pi+t-x}-\sqrt{\pi-t-x})\sin t dt}{[(1-r)^2+4r\cos^2\frac{t}{2}]^2} > 0. \tag{3.6}$$

and

$$I_2 = \int\limits_{x}^{\pi-x} \frac{(\sqrt{t+x}-\sqrt{t-x})\sin t}{[(1-r)^2+4r\sin^2\frac{t}{2}]^2}dt \ge \int\limits_{x}^{2x} \frac{(\sqrt{t+x}-\sqrt{t-x})\sin t}{[(1-r)^2+4r\sin^2\frac{t}{2}]^2}dt$$

$$\ge \int\limits_{x}^{2x} \frac{(\sqrt{t+x}-\sqrt{x})\sin t dt}{[(1-r)^2+4r\sin^2\frac{t}{2}]^2} > \int\limits_{x}^{2x} \frac{\sqrt{x}\sin t dt}{3[(1-r)^2+4r\sin^2\frac{t}{2}]^2}$$

$$> \int\limits_{x}^{2x} \frac{\sqrt{x}\sin t dt}{3[(1-r)^2+4r\sin^2 x]^2} \ge \frac{x\sqrt{x}\sin x}{75(1-r)^4}$$

$$\ge \frac{C(1-r)^2\sqrt{1-r}}{(1-r)^4} = \frac{C}{(1-r)\sqrt{1-r}}. \tag{3.7}$$

Thus, for $\sin x = 1-r$, from (3.4), (3.5), (3.6) and (3.7) we find that

$$\frac{\partial U(\varphi;r,x)}{\partial x} < \frac{C}{\sqrt{1-r}},$$

whence

$$\frac{\partial U(\varphi;r,x)}{\partial x} \to \infty,$$

as $(r,x)\to(1,0)$ along the chosen path.

2. The function $g(x)$ is defined as follows:

$$g(t)=\begin{cases}\sqrt{-t}, & \text{when} \quad -\pi<t\le 0,\\ -\sqrt{t}, & \text{when} \quad 0\le t<\pi,\\ 0, & \text{when} \quad t=\pm\pi\end{cases}$$

and $g(2k\pi+t)=g(t)$, $k=\pm1,\pm2,\dots$. The function $g(t)$ is continuous on the interval $]-\pi;\pi[$, and $g_{(2)}(o)^*=0$. Let $(r,x)\to(1,0)$ so that $1-r=\sin x$ and $0<x<\frac{\pi}{4}$. For this function,

$$\frac{\partial^2 U(g;r,x)}{\partial x^2}=\frac{r(1-r^2)}{\pi}\Big\{\int\limits_{-\pi}^{0}\frac{T(r,t-x)\sqrt{-t}}{[(1-r)^2+4r\sin^2\frac{t-x}{2}]^3}\,dt$$

$$-\int\limits_{0}^{\pi}\frac{T(r,t-x)\sqrt{t}}{[(1-r)^2+4r\sin^2\frac{t-x}{2}]^3}\,dt\Big\}=C(1-r)(I_1-I_2),$$

where

$$T(r,t-x)=4r\sin^2(t-x)-4r\cos(t-x)\sin^2\frac{t-x}{2}-(1-r)^2\cos(t-x),$$

$$I_1=\int\limits_0^{\pi}\frac{T(r,t+x)\sqrt{t}}{(\sin^2x+4rsin^2\frac{t+x}{2})^3}\,dt=\int\limits_x^{\pi+x}\frac{T(r,t)\sqrt{t-x}}{(\sin^2x+4rsin^2\frac{t}{2})^3}\,dt,$$

$$I_2=\int\limits_{-x}^{\pi-x}\frac{T(r,t)\sqrt{t+x}}{(\sin^2x+4rsin^2\frac{t}{2})^3}\,dt,$$

$$I_1-I_2=-\int\limits_x^{\pi-x}\frac{T(r,t)}{(\sin^2x+4rsin^2\frac{t}{2})^3}(\sqrt{t+x}-\sqrt{t-x})dt$$

$$-\int\limits_{-x}^{x}\frac{T(r,t)\sqrt{t+x}}{(\sin^2x+4rsin^2\frac{t}{2})^3}\,dt+\int\limits_{\pi-x}^{\pi+x}\frac{T(r,t)\sqrt{t-x}}{(\sin^2x+4rsin^2\frac{t}{2})^3}\,dt$$

$$=-\int\limits_x^{\pi-x}\frac{T(r,t)(\sqrt{t+x}-\sqrt{t-x})}{(\sin^2x+4rsin^2\frac{t}{2})^3}\,dt-\int\limits_0^{x}\frac{T(r,t)\sqrt{t+x}}{(\sin^2x+4rsin^2\frac{t}{2})^3}\,dt$$

$$-\int\limits_0^{x}\frac{T(r,t)\sqrt{x-t}}{(\sin^2x+4rsin^2\frac{t}{2})^3}\,dt+o(1)=[t=x\tau]$$

$$=-\int\limits_1^{\frac{\pi-x}{x}}\frac{T(r,tx)\sqrt{x}(\sqrt{1+t}-\sqrt{t-1})}{(\sin^2x+4r\sin^2\frac{tx}{2})^3}\,xdt$$

$$-\int_0^1 \frac{T(r,tx)\sqrt{x}\cdot x\sqrt{1+t)}}{(\sin^2 x+4r\sin^2\frac{tx}{2})^3}\,dt-\int_0^1 \frac{T(r,tx)\sqrt{x}\cdot x\sqrt{1-t}}{(\sin^2 x+4r\sin^2\frac{tx}{2})^3}\,dt+o(1).$$

It can be easily verified that

$$\lim_{(r,x)\to(1,0)}\frac{T(r,tx)}{x^2}=\lim_{(r,x)\to(1,0)}\frac{4r\sin^2 tx-4r\cos tx\sin^2\frac{tx}{2}-\sin^2 x\cos tx}{x^2}=3t^2-1,$$

$$\lim_{(r,x)\to(1,0)}\frac{\sin^2 x+4r\sin^2\frac{tx}{2}}{x^2}=1+t^2.$$

Therefore

$$4r\sin^2 tx-4r\cos tx\sin^2\frac{tx}{2}-\sin^2 x\cos tx=x^2(3t^2-1)+o(1)x^2,$$

$$\sin^2 x+4r\sin^2\frac{tx}{2}=(1+t^2)x^2+o(1)x^2.$$

Consequently,

$$\frac{\partial^2 U(g;r,x)}{\partial x^2}=\frac{C(1-r)\sqrt{x}\cdot x^3}{x^6}\Bigg(-\int_1^{\frac{\pi-x}{x}}\frac{3t^2-1}{(1+t^2)^3}(\sqrt{t+1}-\sqrt{t-1})dt$$

$$-\int_0^1\frac{3t^2-1}{(1+t^2)^3}\sqrt{1+t}dt-\int_0^1\frac{3t^2-1}{(1+t^2)^3}\sqrt{1-t}dt\Bigg)+o(1)$$

$$=\frac{C}{x^{3/2}}\Bigg[-\int_1^{\frac{\pi-x}{x}}\frac{3t^2-1}{(1+t^2)^3}(\sqrt{t+1}-\sqrt{t-1})dt-\int_{1/\sqrt{3}}^1\frac{3t^2-1}{(1+t^2)^3}(\sqrt{1+t}+\sqrt{1-t})dt$$

$$+\int_0^{1/\sqrt{3}}\frac{1-3t^2}{(1+t^2)^3}(\sqrt{1+t}+\sqrt{1-t})dt\Bigg]+o(1).$$

whence we obtain (see the proof of Theorem 1.3.1)

$$\frac{\partial^2 U(g;r,x)}{\partial x^2}\to+\infty$$

as $(r,x)\to(1,0)$ along the chosen path.

Theorem 2.3.1 is proved. □

Let $F(x)$ be an undefined integral of the integrable and periodic function f, i.e., $F(x)=\int\limits_{-\pi}^{x}f(t)dt$.

Corollary 2.3.1 ([34], p. 168). *At every point x_0 at which there exists $F^*_{(1)}(x_0)$, we have*

$$\lim_{r\to 1-} U(f;r,x_0) = F^*_{(1)}(x_0).$$

Proof. Assuming without loss of generality that the free term in $S[f]$ is equal to zero, the integration by parts leads to

$$U(f;r,x) = -\frac{1}{\pi}\int_{-\pi}^{\pi} F(t)\frac{\partial P(r,t-x)}{\partial t}\,dt = \frac{\partial}{\partial x}U(F;r,x). \tag{3.8}$$

By virtue of Theorem 2.3.1, this equality shows that Corollary 2.3.1 is valid. □

Theorem 2.3.2. (a) *If at the point x_0 there exists a finite function $f_{(n)}(x_0)$, then*

$$\lim_{(r,x)\overset{\wedge}{\longrightarrow}(1,x_0)} \frac{\partial^n U(f;r,x)}{\partial x^n} = f_{(n)}(x_0).$$

(b) *There exists a function g such that $g'(x_0)$ is finite, but the limit*

$$\lim_{(r,x)\to(1,x_0)} \frac{\partial U(g;r,x)}{\partial x}$$

does not exist.

Proof of Item (a). Let $x_0 = 0$. By Lemma 2.2.2, we may assume that $f(0) = f_{(1)}(0) = \cdots = f_{(n)}(0) = 0$.

Let $\varepsilon > 0$ and choose $\delta > 0$ such that

$$|f(t)| < \varepsilon|t|^n, \quad \text{for} \quad |t| < \delta. \tag{3.9}$$

Then

$$\frac{\partial^n U(f;r,x)}{\partial x^n} = \frac{1}{\pi}\int_{V_\delta} \frac{\partial^n P(r,t-x)}{\partial x^n} f(t)dt + \frac{1}{\pi}\int_{CV_\delta} \frac{\partial^n P(r,t-x)}{\partial x^n} f(t)dt = I_1 + I_2,$$

where $V_\delta = [-\delta;\delta]$.

By virtue of (3.9) and Statement 4) of Lemma 2.2.1, we have

$$|I_1| < \frac{\varepsilon}{\pi}\int_{-\pi}^{\pi}\left|\frac{\partial^n P(r,t-x)}{\partial x^n}\right||t^n|dt = \frac{\varepsilon}{\pi}\int_{-\pi}^{\pi}\left|\frac{\partial^n P(r,t)}{\partial t^n}\right||t+x|^n dt$$

$$< C\varepsilon\int_{-\pi}^{\pi}\left|\frac{\partial^n P(r,t)}{\partial t^n}\right|(|t|^n + |x|^n)dt < C\varepsilon\Big\{\int_{-\pi}^{\pi}\left|\frac{\partial^n P(r,t)}{\partial t^n}\right||t|^n dt$$

$$+\int_{-\pi}^{\pi}\left|\frac{\partial^n P(r,t)}{\partial t^n}\right||x|^n dt\Bigg\} < C\varepsilon + C\varepsilon|x|^n\int_{-\pi}^{\pi}\left|\frac{\partial^n P(r,t)}{\partial t^n}\right|dt$$

$$< C\varepsilon + C\varepsilon|x|^n\int_{-\pi}^{\pi}\frac{I_n(r,t)(1-r)}{[(1-r)^2+4r\sin^2\frac{t}{2}]^{n+1}}dt$$

$$< C\varepsilon + C\varepsilon(1-r)|x|^n\int_{-\pi}^{\pi}\frac{I_n(r,t)dt}{[(1-r)^2+t^2]^{n+1}}dt$$

$$< C\varepsilon + C\varepsilon(1-r)|x|^n\int_{-\frac{\pi}{1-r}}^{\frac{\pi}{1-r}}\frac{(1-r)^{n+1}\left(\sum\limits_{v=0}^{n}C_v|t|^v\right)dt}{(1-r)^{2n+2}(1+t^2)^{n+1}}$$

$$< C\varepsilon + C\varepsilon\frac{|x|^n}{(1-r)^n}\int_0^{\infty}\frac{\left(\sum\limits_{v=0}^{n}C_v t^v\right)dt}{(1+t^2)^{n+1}} < C\varepsilon \tag{3.10}$$

for

$$|t| < \frac{\delta}{2} \quad \text{and} \quad \frac{|x|}{1-r} < C.$$

Next, taking into account Statement 3) of Lemma 2.3.1, we have

$$|I_2| < \max_{\delta\le t\le\pi}\left|\frac{\partial^n P(r,t)}{\partial t^n}\right|\int_{-\pi}^{\pi}|f(t)|dt \quad \text{for} \quad |x| < \frac{\delta}{2}. \tag{3.11}$$

It follows from (3.10) and (3.11) that Item (a) of Theorem 2.3.2 is valid.

Proof of Item (b). Let

$$g(t) = \begin{cases}\sqrt{|t^3|}sin\frac{1}{t}, & \text{when} \quad -\pi < t < \pi \quad \text{and} \quad t \neq 0,\\ 0, & \text{when} \quad t=0 \quad \text{and} \quad t=\pm\pi\end{cases}$$

when $g(2k\pi + t) = g(t)$ and $k = \pm1, \pm2, \ldots$. In Section 1.3 it has been shown that $g'(o) = 0$, but $\overline{g}'(0)$ does not exist. Let $(r, x) \to (1, 0)$ so that $1 - r = \sin^2 x$, $x > 0$. Then for the given function,

$$\frac{\partial U(g;r,x)}{\partial x} = \frac{r(1-r^2)}{\pi}\int_{-\pi}^{\pi}\frac{\sqrt{|t^3|}\sin(t-x)\sin\frac{1}{t}}{[(1-r)^2+4r\sin^2\frac{t-x}{2}]^2}dt$$

$$= \frac{r(1-r^2)}{\pi}\int_{-\pi-x}^{\pi-x}\frac{\sqrt{|t+x|^3}\sin t\sin\frac{1}{t+x}}{[(1-r)^2+4r\sin^2\frac{t}{2}]^2}dt$$

$$= \frac{r(1-r^2)}{\pi} \int_0^{\pi-x} \frac{[\sqrt{(x+t)^3}\sin\frac{1}{t+x} - \sqrt{|x-t|^3}\sin\frac{1}{x-t}]\sin t}{[(1-r)^2+4r\sin^2\frac{t}{2}]^2}\,dt + o(1)$$

$$= C\sin^2 x \int_0^{x} \frac{\left[\sqrt{(x+t)^3}\sin\frac{1}{t+x} - \sqrt{(x-t)^3}\sin\frac{1}{x-t}\right]\sin t}{(\sin^4 x+4r\sin^2\frac{t}{2})^2}\,dt + o(1).$$

Continuing our reasoning like in proving Item (b) of Theorem 1.3.2, we see that the limit

$$\lim \frac{\partial U(g;r,x)}{\partial x}$$

does not exist as $(r,x)\to(1,0)$ along the chosen path.

Theorem 2.3.2 is proved. □

Corollary 2.3.2 ([34], p. 168). *Let $F(x)=\int_{-\pi}^{x} f(t)dt$. At each point x_0 at which $F'(x_0)=f(x_0)$ exists and is finite (therefore almost everywhere),*

$$\lim_{(r,x)\overset{\wedge}{\longrightarrow}(1,x_0)} U(f;r,x)=f(x_0).$$

The validity of this statement follows directly from Theorem 2.3.2 and equality (3.8).

Lemma 2.3.1. *Given the functions $\alpha_i(x)$, $i=0,1,\dots,n$, there exists a function $T_n(x,t)=\sum_{v=0}^{n} a_v(x)e^{ivt}$ such that*

$$T_n(x,o)=\alpha_0(x),\quad \frac{\partial^k T_n(x,o)}{\partial t^k},\quad k=1,2,\dots,n.$$

This lemma is proved in the same manner as the corresponding statement in Lemma 2.2.2.

$T_n(x,t)$ can also be written in the form

$$T_n(x,t)=\alpha_0(x)+\alpha_1(x)t+\alpha_2(x)\frac{t^2}{2!}+\cdots+\alpha_n(x)\frac{t^n}{n!}+\varepsilon_1(x,t)\frac{t^{n+1}}{n!}, \tag{3.12}$$

where $\lim\limits_{t\to 0}\varepsilon_1(x,t)=0$ for any fixed x.

Lemma 2.3.2. *Let $g(t)=\varepsilon(x,t)t^n$. If $\lim\limits_{(x,t)\to(x_0,0)}\varepsilon(x,t)=0$, then*

$$\lim_{(r,x)\to(1,x_0)} \int_{-\pi}^{\pi} \frac{\partial^n P(r,t)}{\partial t^n} g(x,t)dt=0.$$

The proof of the lemma is given by the inequality

$$\left|\int_{-\pi}^{\pi}\frac{\partial^n P(r,t)}{\partial t^n}g(x,t)dt\right| \le \int_{-\pi}^{\pi}\frac{\partial^n P(r,t)}{\partial t^n}|t^n||\varepsilon(x,t)|dt,$$

after applying Statements 1) and 3) of Lemma 2.2.1.

Theorem 2.3.3. *If at the point x_0 there exists a finite function $\overline{f}_{(n)}(x_0)$, then*

$$\lim_{(r,x)\to(1,x_0)}\frac{\partial^n U(f;r,x)}{\partial x^n} = \overline{f}_{(n)}(x_0).$$

Proof. On the strength of the condition of the theorem,

$$f(x+t) = \alpha_0(x) + \alpha_1(x)t + \cdots + \alpha_{n-1}(x)\frac{t^{n-1}}{(n-1)!} + [\overline{f}_{(n)}(x_0) + \varepsilon(x,t)]\frac{t^n}{n!}, \tag{3.13}$$

where $\lim\limits_{(x,t)\to(x_0,0)}\varepsilon(x,t) = 0$.

We construct $T_n(x,t)$ in such a way that the corresponding coefficients in equalities (3.12) and (3.13) be equal to each other for $i = 0, 1, \ldots, n-1$, and $\lim\limits_{x\to x_0}\alpha_n(x) = \overline{f}_{(n)}(x_0) = \frac{\partial^n T_n(x_0,0)}{\partial t^n}$. Then

$$f(x+t) - T_n(x,t) = [\varepsilon(x,t) + \overline{f}_{(n)}(x_0) - \alpha_n(x) - \varepsilon_1(x,t)t]\frac{t^n}{n!} = \varepsilon_2(x,t)\frac{t^n}{n!}, \tag{3.14}$$

where $\lim\limits_{(x,t)\to(x_0,0)}\varepsilon_2(x,t) = 0$.

Further,

$$\frac{\partial^n U(f;r,x)}{\partial x^n} = \frac{1}{\pi}\int_{-\pi}^{\pi}\frac{\partial^n P(r,t-x)}{\partial x^n}f(t)dt = \frac{(-1)^n}{\pi}\int_{-\pi}^{\pi}\frac{\partial^n P(r,t)}{\partial t^n}f(x+t)dt$$

$$= \frac{(-1)^n}{\pi}\int_{-\pi}^{\pi}\frac{\partial^n P(r,t)}{\partial t^n}[f(x+t) - T_n(x,t)]dt + \frac{(-1)^n}{\pi}\int_{-\pi}^{\pi}\frac{\partial^n P(r,t)}{\partial t^n}T_n(x,t)dt$$

$$= I_1 + I_2, \tag{3.15}$$

where with regard for (3.14),

$$I_1 = \frac{(-1)^n}{\pi n!}\int_{-\pi}^{\pi}\frac{\partial^n P(r,t)}{\partial t^n}\varepsilon_2(x,t)t^n dt.$$

By Lemma 2.3.2, we have

$$\lim_{(r,x)\to(1,x_0)} I_1 = 0. \tag{3.16}$$

It can be easily verified that

$$I_2 = \frac{\partial^n}{\partial t^n}\left(\sum_{v=0}^{n} a_v(x) r^v e^{ivt}\right)\Bigg|_{t=0}.$$

Hence we obtain

$$\lim_{(r,x)\to(1,x_0)} I_2 = \frac{\partial^n T_n(x_0,0)}{\partial t^n} = \overline{f}_{(n)}(x_0). \tag{3.17}$$

From (3.15), (3.16) and (3.17) it follows that our theorem is valid.
Theorem 2.3.3 is proved. □

Corollary 2.3.3. *Let $F(x) = \int_{-\pi}^{x} f(t)dt$. At every point, at which $\overline{F}_{(1)}(x_0) = f(x_0)$ exists and is finite (therefore, by Lemma 1.3.1, for a continuous everywhere function), we have*

$$\lim_{(r,x)\to(1,x_0)} U(f;r,x) = f(x_0).$$

The validity of this statement follows from the equality (3.8) and Theorem 2.3.3.

2.4 The Dirichlet Problem for a Circle

The Dirichlet problem for a unit circle (of radius 1 and with center at the origin) is formulated as follows: given the function $f(t)$ on the circumference $R = 1$, find the function $U(r,x)$, harmonic inside the circle and tending to the prescribed values of $f(\overline{x})$ on the circumference, when the point (r,x) $(r < 1)$ tends in this manner or another to the points of the circumference $(1,\overline{x})$.

Let us give the well-known theorems in terms of the Dirichlet problem which follow from the results of Section 2.3.

Theorem 2.4.1. *If the 2π-periodic function $f(t)$ is continuous, then the Poisson integral $U(f;r,x)$, harmonic inside the circle, is a solution of the Dirichlet problem in the sense that for all $\overline{x} \in [-\pi,\pi]$,*

$$\lim_{(r,x)\to(1,\overline{x})} U(f;r,x) = f(\overline{x})$$

(see Corollary 2.3.3*).*

Theorem 2.4.2. *If the 2π-periodic function $f(t) \in L[-\pi,\pi]$, then the Poisson integral $U(f;t,x)$, harmonic inside the circle, is a solution of the Dirichlet problem in the sense that almost for all points $\overline{x} \in [-\pi,\pi]$,*

$$\lim_{(r,x)\overset{\wedge}{\longrightarrow}(1,\overline{x})} U(f;r,x) = f(\overline{x})$$

(see Corollary 2.3*, the Fatou theorem).*

Using the Fatou theorem and the theorems on the existence of a primitive function, N. N. Luzin ([43], p. 87) solved this Dirichlet problem in the case, where the function $f(t)$ is measurable and finite almost everywhere on $[-\pi, \pi]$.

Theorem 2.4.3. *Let $f(t)$ be an arbitrary measurable and almost everywhere finite function on the circumference $R = 1$. Then inside the circle there exists a harmonic function $U(r, x)$ which is a solution of the Dirichlet problem in the sense that*

$$\lim_{(r,x)\xrightarrow{\wedge}(1,\overline{x})} U(r, x) = f(\overline{x})$$

almost everywhere on the circumference.

Proof. By Luzin's theorem ([43], p. 78) on the existence of a primitive function, there exists for $f(t)$ a continuous function $F(t)$ such that $F'(x) = f(t)$ almost everywhere.

Consider the function

$$U(r, x) = \frac{1}{\pi}\int_{-\pi}^{\pi} F(t)\frac{\partial}{\partial x}\left[\frac{1-r^2}{1-2r\cos(t-x)+r^2}\right]dt.$$

It is clear that the function $U(r, x)$ is harmonic inside the circle. By the Fatou theorem (see Corollary 2.3.2), for the points $\overline{x}$ at which $F'(\overline{x})$ exists, we have

$$\lim_{(r,x)\xrightarrow{\wedge}(1,\overline{x})} U(r, x) = F'(\overline{x}).$$

But since $F'(\overline{x}) = f(\overline{x})$ almost everywhere, Theorem 2.4.3 is proved. □

Chapter 3

Boundary Properties of Derivatives of the Poisson Integral for a Ball

3.1 Notation, Definitions and Statement of Some Well-Known Facts

1. R^k is a k-dimensional Euclidean space; $x = (x_1, x_2, \dots, x_k)$, $t = (t_1, t_2, \dots, t_k)$, $x^0 = (x_1^0, x_2^0, \dots, x_k^0)$ are points of the space R^k; $(x,t) = \sum_{i=1}^{k} x_i t_i$ is a scalar product; $|x| = \sqrt{(x,x)}$; $ht = (ht_1, ht_2, \dots, ht_k)$; $x + t = (x_1 + t_1, x_2 + t_2, \dots, x_k + t_k)$.

2. $S_\rho^{k-1}(x)$ is a $(k-1)$-dimensional sphere, $V_\rho^k(x)$ is a k-dimensional ball in R^k with center at x and of radius ρ ($S_\rho^{k-1} = S_\rho^{k-1}(0)$, $S^{k-1} = S_1^{k-1}$ is the unit sphere; $V_\rho^k = V_\rho^k(0)$, $V^k = V_1^k$); $|S_\rho^{k-1}(x)|$ is a $(k-1)$-dimensional space of the sphere $S_\rho^{k-1}(x)$, and $|S_\rho^{k-1}| = \dfrac{2\pi^{k/2}\rho^{k-1}}{\Gamma(k/2)}$;

$$D^{k-1}(x;h) = \{t \in S^{k-1} : (x,t) > \cos h,\quad 0 < h \leq \pi\};$$

$$C^{k-2}(x;h) = \{t \in S^{k-1} : (x,t) = \cos h,\quad 0 < h \leq \pi\};$$

$$|C^{k-2}(x;h)| = |S^{k-2}| \sin^{k-2} h = \frac{2\pi^{\frac{k-1}{2}} \sin^{k-2} h}{\Gamma\left(\frac{k-1}{2}\right)};$$

$$|D^{k-1}(x;h)| = \int_0^h |C^{k-2}(x,\gamma)| d\gamma = |S^{k-2}| \int_0^h \sin^{k-2} \gamma d\gamma;$$

$|V_\rho^k| = \dfrac{2\pi^{k/2}\rho^k}{k\Gamma(k/2)}$ is the volume of the ball V_ρ^k; $\Gamma(\alpha) = \int_0^\infty e^{-t} t^{\alpha-1} dt$ ($\alpha > 0$) is a gamma-function or a Euler integral of second kind.

3. If $(\rho, \theta_1, \theta_2, \dots, \theta_{k-2}, \varphi)$ are the spherical coordinates of the point $x(x_1, x_2, \dots, x_k)$, then

$$\begin{gathered} x_1 = \rho\cos\theta_1; \\ x_2 = \rho\sin\theta_1\cos\theta_2; \\ x_3 = \rho\sin\theta_1\sin\theta_2\cos\theta_3; \\ \dots\dots\dots\dots\dots\dots \\ x_{k-1} = \rho\sin\theta_1\sin\theta_2\dots\sin\theta_{k-2}\cos\varphi; \\ x_k = \rho\sin\theta_1\sin\theta_2\dots\sin\theta_{k-2}\sin\varphi; \\ 0 \le \rho < \infty; \quad 0 \le \theta_1 \le \pi \quad (i = \overline{1, k-2}), \quad 0 \le \varphi \le 2\pi, \\ dS_\rho^{k-1}(x) = \rho^{k-1}\sin^{k-2}\theta_1\dots\sin\theta_{k-2}d\theta_1\dots d\theta_{k-2}d\varphi \end{gathered} \tag{1.1}$$

is an element of a $(k-1)$-dimensional area of the sphere $S_\rho^{k-1}(dS_1^{k-1} = dS^{k-1})$.

4. If γ is an angle between the radii drawn from the center of the ball $V^k = |x| < 1$ to the points $x(\theta_1, \theta_2, \dots, \theta_{k-2}, \varphi) \in S^{k-1}$, $y(\theta'_1, \theta'_2, \dots, \theta'_{k-2}, \varphi') \in S^{k-1}$, then

$$\begin{aligned} (x, y) = \cos\gamma &= \cos\theta_1\cos\theta'_1 + \sin\theta_1\cos\theta_2\sin\theta'_1\cos\theta'_2 + \cdots \\ &+ \sin\theta_1\sin\theta_2\dots\sin\theta_{k-2}\cos\varphi\sin\theta'_1\sin\theta'_2\dots\sin\theta'_{k-2}\cos\varphi' \\ &+ \sin\theta_1\sin\theta_2\dots\sin\theta_{k-2}\sin\varphi\sin\theta'_1\sin\theta'_2\dots\sin\theta'_{k-2}\sin\varphi'. \end{aligned} \tag{1.2}$$

If $\theta_1 = 0$, which means that x coincides with the pole, then from (1.2) we obtain $\cos\gamma = \cos\theta'_1$ and therefore $\gamma = \theta'_1$. Thus, transforming the coordinate system so that the pole would coincide with x, we have $\theta'_1 = \gamma$.

5. $\Delta = \sum\limits_{v=1}^{k} \dfrac{\partial^2}{\partial x_v^2}$ is a Laplace operator in R^k, $\Delta^r = \Delta(\Delta^{r-1})$; $\Delta' = \Delta$; D_k is a Laplace operator on S^{k-1}, i.e., the angular part of the Laplace operator Δ written in terms of spherical coordinates ([41], p.240; [74], p.59; [4], p.227),

$$\begin{gathered} \Delta = \frac{1}{\rho^{k-1}}\cdot\frac{\partial}{\partial\rho}\Big(\rho^{k-1}\frac{\partial}{\partial\rho}\Big) + \frac{1}{\rho^2}D_k, \\ D_k^r = D_k(D_k^{r-1}), \quad D_k^1 = D_k; \end{gathered} \tag{1.3}$$

$$\begin{aligned} D_k = &\frac{1}{\sin^{k-2}\theta_1}\cdot\frac{\partial}{\partial\theta_1}\Big(\sin^{k-2}\theta_1\frac{\partial}{\partial\theta_1}\Big) \\ &+\frac{1}{\sin^2\theta_1\sin^{k-3}\theta_2}\cdot\frac{\partial}{\partial\theta_2}\Big(\sin^{k-3}\theta_2\frac{\partial}{\partial\theta_2}\Big) \\ &+\frac{1}{\sin^2\theta_1\sin^2\theta_2\sin^{k-4}\theta_3}\cdot\frac{\partial}{\partial\theta_3}\Big(\sin^{k-4}\theta_3\frac{\partial}{\partial\theta_3}\Big) \\ &+\cdot\ \dots\dots\dots\dots\dots\dots \\ &+\frac{1}{\sin^2\theta_1\sin^2\theta_2\dots\sin^2\theta_{k-3}\sin\theta_{k-2}}\cdot\frac{\partial}{\partial\theta_{k-2}}\Big(\sin\theta_{k-2}\frac{\partial}{\partial\theta_{k-2}}\Big) \end{aligned}$$

$$+\frac{1}{\sin^2\theta_1 \sin^2\theta_2 \ldots \sin^2\theta_{k-2}} \cdot \frac{\partial^2}{\partial\varphi^2}. \tag{1.4}$$

It follows from (1.3) that if $U(x) = U(\rho, \theta_1, \theta_2, \ldots, \theta_{k-2}, \varphi)$ is a harmonic function satisfying the Laplace equation $\sum\limits_{v=1}^{k} \frac{\partial^2 U(x)}{\partial x_v^2} = 0$, then

$$D_k U(x) = -\frac{1}{\rho^{k-3}} \cdot \frac{\partial}{\partial\rho}\Big(\rho^{k-1}\frac{\partial U(x)}{\partial\rho}\Big). \tag{1.5}$$

6. If for $k = 3$, (x, y, z) are the Cartesian coordinates of the point M, and (ρ, θ, φ) are the spherical coordinates, then the equalities (1.1),(1.3) and (1.4) take the form

$$x = \rho\sin\theta\cos\varphi, \quad y = \rho\sin\theta\sin\varphi, \quad z = \rho\cos\theta,$$
$$dS_\rho^2 = \rho^2 \cdot \frac{\partial}{\partial\rho}\Big(\rho^2\frac{\partial}{\partial\rho}\Big) + \frac{1}{\rho^2}D_3,$$
$$D_3 = \frac{1}{\sin\theta}\frac{\partial}{\partial\theta}\Big(sin\theta\frac{\partial}{\partial\theta}\Big) + \frac{1}{\sin^2\theta} \cdot \frac{\partial^2}{\partial\varphi^2}.$$

7. $L_p(S^{k-1})$, $1 \le p < \infty$ $(L_1(S^{k-1}) = L(S^{k-1}))$ is the space of the function f with the norm

$$\|f\|_{L_P(S^{k-1})} = \Big(\int\limits_{S^{k-1}} |f(x)|^P dS^{k-1}(x)\Big)^{1/P}, \quad 1 \le P < \infty.$$

For $P = \infty$ it is assumed that the space $L_\infty(S^{k-1}) = C(S^{k-1})$ consists of continuous functions with the norm

$$\|f\|_{C(S^{k-1})} = \max_{x\in S^{k-1}} |f(x)|.$$

$M(S^{k-1})$ is the space of finite, regular Borel measures μ with the norm

$$\|\mu\|_{M(S^{k-1})} = \int\limits_{S^{k-1}} |d\mu(x)|.$$

8. $(f, g) = \int\limits_{S^{k-1}} f(x)g(x)dS^{k-1}(x)$ is the scalar product of real functions of the class $L_2(S^{k-1})$.

9. Let P_n be a set of all homogeneous polynomials of degree n defined on R^k. If $P \in P_n$, then ([66], p. 158)

$$P(x) = \sum_{|\alpha|=n} C_\alpha x^\alpha,$$

where $\alpha = (\alpha_1, \alpha_2, \ldots, \alpha_k)$ is the multi-index (of nonnegative integers), $|\alpha| = \alpha_1 + \alpha_2 + \cdots + \alpha_k$ and $x^\alpha = x_1^{\alpha_1} x_2^{\alpha_2} \ldots x_k^{\alpha_k}$. Clearly, the monomials x^n, $|\alpha| = n$ form the basis of that space. The number of such monomials is equal to the number $d_n = \dim P_n$ of various multi-indices $\alpha = (\alpha_1, \alpha_2, \cdots, \alpha_k)$ such that $\alpha_1 + \alpha_2 + \cdots + \alpha_k = n$ and calculated by the formula

$$d_n = \binom{k+n-1}{k-1} = \binom{k+n-1}{n} = \frac{(k+n-1)!}{(k-1)!n!}.$$

10. A_n is the class of all harmonic polynomials from P_n. A_n is also called the space of spatial spherical harmonics.

11. The restriction of a homogeneous harmonic polynomial of degree n to the unit sphere S^{k-1} is called a surface spherical harmonic of degree n or a spherical harmonic of degree n.

12. H_n is the space of spherical harmonics of degree n. It coincides ([66], p. 160) with the set of restrictions of all elements from A_n to S^{k-1}. The dimension of H_n is equal to

$$a_n = \dim H_n = \dim A_n = d_n - d_{n-2} = (2n+k-2)\frac{(n+k-3)!}{n!(k-2)!}. \tag{1.6}$$

Note that the restriction of any polynomial of k variables to the unit sphere S^{k-1} is the sum of restrictions of harmonic polynomials to the unit sphere S^{k-1} ([66], p. 160; [59], p. 423).

13. Let Y_n and Y_m be spherical harmonics of degrees n and m, respectively, where $n \neq m$. Then ([66], p. 161)

$$\int\limits_{S^{k-1}} Y_n(x) Y_m(x) dS^{k-1}(x) = 0.$$

14. The set of spherical harmonics of degree n can be considered as a subspace of the space $L_2(S^{k-1})$ of real functions with the scalar product (f, g). If $\{Y_1^{(n)}, Y_2^{(n)}, \ldots, Y_{a_n}^{(n)}\}$ is the orthonormalized basis of this subspace, then the set of functions

$$\bigcup_{n=0}^{\infty} \{Y_1^{(n)}, Y_2^{(n)}, \ldots, Y_{a_n}^{(n)}\}$$

is ([66], p. 161) the orthonormalized basis in the space $L_2(S^{k-1})$.

15. The addition theorem

$$P_n^\lambda[(x, y)] = \frac{2\pi^{\lambda+1}}{(n+\lambda)\Gamma(\lambda)} \sum_{j=1}^{a_n} Y_j^{(n)}(x) Y_j^{(n)}(y), \quad 2\lambda = k-2, \tag{1.7}$$

is valid ([5], p. 206); here $P_n^\lambda(t)$ are Gegenbauer polynomials ([3], p. 177; [112], p. 149) (ultraspherical polynomials) defined by the decomposition ([15], p. 247; [5], p. 205; [58], p. 94)

$$(1-2th+h^2)^{-\lambda} = \sum_{n=0}^{\infty} P_n^\lambda(t)h^n \tag{1.8}$$

or

$$\frac{1-h^2}{(1-2th+h^2)^{\lambda+1}} = \sum_{n=0}^{\infty} \frac{n+\lambda}{\lambda} P_n^\lambda(t)h^n \tag{1.9}$$

for $0 \le h < 1$ and $\lambda > -\frac{1}{2}$, $\lambda \ne 0$. For $\lambda = 0$, by virtue of

$$\lim_{\lambda\to 0} \lambda^{-1} P_n^\lambda(\cos\theta) = \frac{2}{n}\cos n\theta \quad (n \ge 1, \quad t = \cos\theta),$$

the formula

$$\frac{1-h^2}{1-2h\cos\theta+h^2} = 1 + 2\sum_{n=1}^{\infty} h^n \cos n\theta$$

is valid.

For $\lambda = \frac{1}{2}$, the Gegenbauer polynomials are Legendre polynomials $P_n(t)$.

Gegenbauer polynomials $P_n^\lambda(t)$ can also be obtained by orthogonalization of functions $1, t, t^2 \ldots$ on the interval $[-1, 1]$ with weight $(1-t^2)^{\lambda-\frac{1}{2}}$. Note that

$$P_n^\lambda(1) = \frac{\Gamma(n+2\lambda)}{n!\Gamma(2\lambda)} = \binom{n+2\lambda-1}{n}, \quad n = 0, 1, 2, \ldots$$

16. If $Y_n(x) \in H_n$, then

$$Y_n(y) = \frac{\Gamma(\lambda)(n+\lambda)}{2\pi^{\lambda+1}} \int\limits_{S^{k-1}} Y_n(x) P_n^\lambda[(x,y)]dS^{k-1}(x).$$

17. If $Y_n(x) \in H_n$, then ([5], p. 205; [65], p. 86)

$$D_k Y_n(x) = -n(n+k-2)Y_n(x). \tag{1.10}$$

18. It follows from (1.6) that in the space R^3 there exist $(2n+1)$ linearly independent spherical harmonics of degree n. These functions can be written in the explicit form

$$\begin{gathered} P_n(\cos\theta); \quad P_n^m(\cos\theta)\cos m\varphi; \quad P_n^m(\cos\theta)\sin m\varphi, \\ 0 \le \theta \le \pi, \quad 0 \le \varphi \le 2\pi \quad (m = 1, 2, \ldots, n; \quad n = 0, 1, 2, \ldots) \end{gathered} \tag{1.11}$$

where $P_n(t)$ are ordinary, and $P_n^m(t)$ adjoint Legendre polynomials

$$P_n^m(\cos\theta) = \sin^m\theta \frac{d^m P_n(\cos\theta)}{(d\cos\theta)^m}.$$

The functions (1.11) are particular solutions of the equation

$$\frac{1}{\sin\theta}\cdot\frac{\partial}{\partial\theta}\Big(\sin\theta\frac{\partial Y}{\partial\theta}\Big)+\frac{1}{\sin^2\theta}\cdot\frac{\partial^2 Y}{\varphi^2}+n(n+1)Y=0.$$

The functions (1.11) are linearly independent and therefore any spherical function $Y_n(\theta,\varphi)$ of order n can be written in the form of their linear combination

$$Y_n(\theta,\varphi)=\frac{1}{2}\alpha_{n,0}P_n(\cos\theta)$$

$$+\sum_{m=1}^{n}P_n^m(\cos\theta)(\alpha_{n,m}\cos m\varphi+\beta_{n,m}\sin m\varphi).$$

The functions (1.11) are mutually orthogonal on the sphere S^2 and, moreover,

$$\|(P_n(\cos\theta)\|^2_{L_2(S^2)}=\frac{4\pi}{2n+1},$$

$$\|(P_n^m(\cos\theta)\cos m\varphi\|^2_{L_2(S^2)}=\|(P_n^m(\cos\theta)\sin m\varphi\|^2_{L_2(S^2)}=\frac{2\pi}{2n+1}\cdot\frac{(n+m)!}{(n-m)!}.$$

19. A series of the form

$$\sum_{j=0}^{\infty}\Big[\frac{1}{2}\alpha_{j,0}P_j(\cos\theta)+\sum_{m=1}^{j}P_j(\cos\theta)(\alpha_{j,m}\cos m\varphi+\beta_{j,m}\sin m\varphi)\Big], \tag{1.12}$$

where $\alpha_{j,0}$, $\alpha_{j,m}$ and $\beta_{j,m}$ are constants, is called a Laplace series in the space R^3.

The coefficients $\alpha_{j,m}$ and $\beta_{j,m}$ are given for $j\geq 0$, $0\leq m\leq j$ and $j\geq 1$, $1\leq m\leq j$, respectively. The definitions of $\alpha_{j,m}$ and $\beta_{j,m}$ can be extended to the rest of the integers m if we assume that

$$\alpha_{j,-m}=\alpha_{j,m}\ (m>0),\quad \beta_{j,0}=0,\quad \beta_{j,-m}=-\beta_{j,m}\ (m>0)$$

and denote

$$\gamma_{j,m}=\frac{1}{2}(\alpha_{j,m}-i\beta_{j,m}).$$

Thus

$$\left.\begin{aligned}\alpha_{j,m}&=\gamma_{j,m}+\gamma_{j,-m}\\ \beta_{j,m}&=i(\gamma_{j,m}-\gamma_{j,-m}\end{aligned}\right\}. \tag{1.13}$$

Conversely, given $\gamma_{j,m}$, we can define $\alpha_{j,m}$ and $\beta_{j,m}$ from (1.13). Then

$$Y_j(\theta,\varphi)=\frac{1}{2}\alpha_{j,0}P_j(cos\theta)+\sum_{m=1}^{j}P_j^m(\cos\theta)(\alpha_{j,m}\cos m\varphi+\beta_{j,m}\sin m\varphi)$$

$$=\gamma_{j,0}P_j(\cos\theta)+\sum_{m=1}^{j}P_j^m(\cos\theta)[(\gamma_{j,m}+\gamma_{j,-m})\cos m\varphi$$

$$+i(\gamma_{j,m}-\gamma_{j,-m})\sin m\varphi]$$

$$=\gamma_{j,0}P_j(\cos\theta)+\sum_{m=1}^{j}P_j^m(\cos\theta)[\gamma_{j,m}(\cos m\varphi+i\sin m\varphi)$$

$$+\gamma_{j,-m}(\cos m\varphi-i\sin m\varphi)]=$$

$$\gamma_{j,0}P_j(\cos\theta)+\sum_{m=1}^{j}P_j^m(\cos\theta)(\gamma_{j,m}e^{im\varphi}+\gamma_{j,-m}e^{-im\varphi})$$

$$=\sum_{m=-j}^{j}\gamma_{j,m}P_j^{|m|}(\cos\theta)e^{im\varphi};$$

$$P_j^0(\cos\theta)=P_j(\cos\theta).$$

The series (1.12) now takes the form

$$\sum_{j=0}^{\infty}\sum_{m=-j}^{j}\gamma_{j,m}P_j^{|m|}(\cos\theta)e^{im\varphi}. \tag{1.14}$$

We call (1.12) areal, and (1.14) a complex Laplace series.

20. Let $f\in L(S^2)$ (if some function f is given on the unit sphere, then this function is regarded as extended to the whole space except for zero and infinity, but remaining constant on the rays emanating from the origin).

The series (1.12) is called the Fourier–Laplace series of the function f if

$$\left.\begin{aligned}
&\alpha_{j,m}=\tfrac{2j+1}{2\pi}\cdot\tfrac{(j-m)!}{(j+m)!}\int_0^{2\pi}\cos m\varphi d\varphi\int_0^{\pi}f(\theta,\varphi)\\
&\times P_j^m(\cos\theta)\sin\theta d\theta,\quad j=0,1,2,\ldots;\quad m=0,1,2,\ldots,j;\\
&\beta_{j,m}=\tfrac{2j+1}{2\pi}\cdot\tfrac{(j-m)!}{(j+m)!}\int_0^{2\pi}\sin m\varphi d\varphi\int_0^{\pi}f(\theta,\varphi)\\
&\times P_j^m(\cos\theta)\sin\theta d\theta,\quad j=1,2,\ldots;\quad m=1,2,\ldots,j;
\end{aligned}\right\} \tag{1.15}$$

$$\gamma_{j,m}=\overline{\gamma}_{j,-m}=\frac{1}{2}(\alpha_{j,m}-i\beta_{j,m})$$

$$=\frac{2j+1}{4\pi}\cdot\frac{(j-|m|)!}{(j+|m|)!}\int_0^{2\pi}e^{-im\varphi}d\varphi\int_0^{\pi}f(\theta,\varphi)P_j^{|m|}(\cos\theta)\sin\theta d\theta,$$

$$j=0,1,2,\ldots;\quad m=0,\pm1\pm2,\ldots,\pm j.$$

If $\alpha_{j,m}$ and $\beta_{j,m}$ are defined by the equalities (1.15), then using the addition formula, we obtain ([9], p. 143)

$$Y_j(\theta,\varphi)=\frac{2j+1}{4\pi}\int_0^{\pi}\int_0^{2\pi}f(\theta',\varphi')P_j(\cos\gamma)\sin\theta' d\theta' d\varphi',$$

where

$$\cos\gamma = \cos\theta\cos\theta' + \sin\theta\sin\theta'\cos(\varphi - \varphi').$$

The Fourier–Laplace series of the function $f \in L(S^2)$ is considered in the form

$$f(\theta,\varphi) \sim \frac{1}{4\pi}\sum_{j=0}^{\infty}(2j+1)\int_0^{\pi}\int_0^{2\pi} f(\theta',\varphi')P_j(\cos\gamma)\sin\theta' d\theta' d\varphi'.$$

21. Let us now introduce the notion of a Fourier–Laplace series in a space R^k, $k > 3$.

Let $f \in L(S^{k-1})$. Its Fourier–Laplace series is

$$S(f;x) = \sum_{n=0}^{\infty} Y_n^{\lambda}(f;x), \tag{1.16}$$

where

$$Y_n^{\lambda}(f;x) = b_1^{(n)}Y_1^{(n)}(x) + b_2^{(n)}Y_2^{(n)}(x) + \cdots + b_{a_n}^{(n)}Y_{a_n}^{(n)}(x), \tag{1.17}$$
$$b_j^{(n)} = (f, Y_j^{(n)}), \quad j = 1,2,\dots,a_n.$$

Taking (1.7) into account, from (1.17) we obtain

$$Y_n^{\lambda}(f;x) = \frac{(n+\lambda)\Gamma(\lambda)}{2\pi^{\lambda+1}}\int_{S^{k-1}} P_n^{\lambda}[(x,y)]f(y)dS^{k-1}(y) \tag{1.18}$$

([5], p. 206).

22. Let $\mu \in M(S^{k-1})$. As is known ([57], p. 176), μ has, at almost all points $x \in S^{k-1}$, a derivative $\mu_S(x)$ under which we mean the limit

$$\mu_S(x) = \lim_{r\to 0}\frac{\mu[D^{k-1}(x;r)]}{|D^{k-1}(x;r)|}. \tag{1.19}$$

A Fourier–Laplace–Stieltjes series is defined as follows:

$$S(d\mu;x) = \sum_{n=0}^{\infty} Y_n^{\lambda}(d\mu;x), \tag{1.20}$$

where

$$Y_n^{\lambda}(d\mu;x) = \frac{(n+\lambda)\Gamma(\lambda)}{2\pi^{\lambda+1}}\int_{S^{k-1}} P_n^{\lambda}[(x,y)]d\mu(y)$$

([5], p. 208; [68], p. 515).

If μ is absolutely continuous, then

$$S(d\mu;x) = S(\mu_s;x).$$

23. If $f \in L_2(S^{k-1})$, then we have the Parseval equality

$$\int\limits_{S^{k-1}} f^2(x)dS^{k-1}(x) = \sum_{n=0}^{\infty} \int\limits_{S^{k-1}} [Y_n^\lambda(f;x)]^2 dS^{k-1}(x)$$

([65], p. 86).

3.2 The Poisson Integral for a Ball

Let $f \in L_2(S^{k-1})$, and

$$S(f;x) = \sum_{n=0}^{\infty} Y_n^\lambda(f;x), \quad x \in S^{k-1}, \tag{2.1}$$

be its Fourier–Laplace series (see (1.16) and (1.18)). The abelian means of the series (2.1) are denoted by

$$U(f;\rho,x) = U(f;\rho,\theta_1,\theta_2,\dots,\theta_{k-2},\varphi) = \sum_{n=0}^{\infty} Y_n^\lambda(f;x)\rho^n, \quad x \in S^{k-1}.$$

But (see (1.18))

$$Y_n^\lambda(f;x) = \frac{\Gamma(\lambda)(n+\lambda)}{2\pi^{\lambda+1}} \int\limits_{S^{k-1}} P_n^\lambda(\cos\gamma)f(y)dS^{k-1}(y), \quad n = \overline{0,\infty},$$

where $\cos\gamma$ is defined by the equality (1.2). Therefore

$$U(f;\rho,x) = \frac{\Gamma(\lambda)}{2\pi^{\lambda+1}} \sum_{n=0}^{\infty}(n+\lambda)\rho^n \int\limits_{S^{k-1}} P_n^\lambda(\cos\gamma)f(y)dS^{k-1}(y).$$

Since $0 < \rho < 1$, by virtue of $|P_n^\lambda(\cos\gamma)| \le Cn^{2\lambda-1}$ ([58], p. 197) the series $\sum_{n=0}^{\infty} P_n^\lambda(\cos\gamma)\rho^n$ converges uniformly with respect to γ for fixed ρ; therefore, after multiplying by $f(y)$, it can be integrated termwise. As a result,

$$\begin{aligned} U(f;\rho,x) &= \frac{\Gamma(\lambda)}{2\pi^{\lambda+1}} \int\limits_{S^{k-1}} \left\{ \sum_{n=0}^{\infty} (n+\lambda)P_n^\lambda(\cos\gamma)\rho^n \right\} f(y)dS^{k-1}(y) \\ &= \frac{\Gamma(\lambda)}{2\pi^{\lambda+1}} \int\limits_{S^{k-1}} P(\rho,\gamma)f(y)dS^{k-1}(y) \\ &= \frac{\Gamma(\lambda+1)}{2\pi^{\lambda+1}} \int\limits_{S^{k-1}} P(\rho,\gamma)f(y)dS^{k-1}(y), \end{aligned} \tag{2.2}$$

where

$$P(r,\gamma)=\sum_{n=0}^{\infty}\frac{n+\lambda}{\lambda}P_n^{\lambda}(\cos\gamma)\rho^n. \tag{2.3}$$

The integral (2.2) is called the Poisson integral of the function f for a ball, and $P(\rho,\gamma)$ is called the Poisson kernel. Therefore the expressions "abelian means of the Fourier–Laplace series of the function f" and "the Poisson integral of the function f for a ball" are synonyms.

Let us find a simpler expression for the Poisson kernel, i.e., for the series (2.3). By the definition of $P_n^{\lambda}(\cos\gamma)$ (see (1.8)),

$$\frac{1}{(1-2\rho\cos\gamma+\rho^2)^{\lambda}}=\sum_{n=0}^{\infty}\rho^n P_n^{\lambda}(\cos\gamma). \tag{2.4}$$

Differentiating the series (2.3) with respect to ρ, we obtain

$$\frac{2\lambda\rho(\cos\gamma-\rho)}{(1-2\rho\cos\gamma+\rho^2)^{\lambda+1}}=\sum_{n=1}^{\infty}n\rho^n P_n^{\lambda}(\cos\gamma). \tag{2.5}$$

By (2.4) and (2.5) we conclude that

$$P(\rho,\gamma)=\sum_{n=0}^{\infty}\frac{n+\lambda}{\lambda}P_n^{\lambda}(\cos\gamma)\rho^n=\frac{1-\rho^2}{(1-2\rho\cos\gamma+\rho^2)^{\lambda+1}}$$
$$=\frac{1-\rho^2}{[(1-\rho)^2+4\rho\sin^2\frac{\gamma}{2})]^{\lambda+1}}.$$

The symbol $(\rho,x)\xrightarrow{\wedge}(1,x^0)$ denotes (see Section 2.1) that the point $(\rho,x)=(\rho,\theta_1,\theta_2,\dots,\theta_{k-2},\varphi)$ tends to the point $(1,x^0)=(1,\theta_1^0,\theta_2^0,\dots,\theta_{k-2}^0,\varphi^0)$ along a nontangential path to the sphere S^{k-1}. This means that there exists a positive constant C such that $\dfrac{\rho_0}{1-\rho}<C$, where ρ_0 is a distance between the points (ρ,x) and $(1,x^0)$.

The symbol $(\rho,x)\to(1,x^0)$ means that the point (ρ,x) tends to $(1,x^0)$ arbitrarily, remaining inside the unit sphere S^{k-1}.

The series (2.1) is called summable by the Abel method at the point $x^0(1,\theta_1^0,\theta_2^0,\dots,\theta_{k-2}^0,\varphi^0)$ to the number S (or, briefly, A-summable to the number S) if

$$\lim_{\rho\to1-}U(f;\rho,x^0)=S.$$

Further, the series (2.1) is called summable by the method A^* at the point x^0 to the number S if

$$\lim_{(\rho,x)\xrightarrow{\wedge}(1,x^0)}U(f;\rho,x)=S.$$

The expressions "summability of the Fourier–Laplace series of the function f by the Abel method" and "the boundary properties of the Poisson integral of the function f for a ball" are synonyms. The following theorems are well-known.

Theorem 3.2.1 ([95], p. 107). *If the function $f(x)$ is continuous at a point $x^0 \in S^{k-1}$, then*

$$\lim_{(\rho,x)\to(1,x^0)} U(f;\rho,x) = f(x^0).$$

Theorem 3.2.2 ([95], p. 110). *If $f \in L(S^{k-1})$, then almost for all points $\overline{x} \in S^{k-1}$*

$$\lim_{(\rho,x)\overset{\wedge}{\longrightarrow}(1,\overline{x})} U(f;\rho,x) = f(\overline{x}).$$

Theorem 3.2.3 ([95], p. . 111). *If for $f \in L(S^{k-1})$,*

$$\Psi(e) = \int_e f(y)dS^{k-1}(y),$$

then for almost all points $\overline{x} \in S^{k-1}$,

$$\lim_{(\rho,x)\overset{\wedge}{\longrightarrow}(1,\overline{x})} U(d\Psi;\rho,x) = \Psi_S(\overline{x}) = f(\overline{x}),$$

where (see (1.19))

$$\Psi_S(\overline{x}) = \lim_{\rho\to 0} \frac{\Psi[D^{k-1}(\overline{x},\rho)]}{D^{k-1}(\overline{x},\rho)}.$$

3.3 A Generalized Laplace Operator on the Unit Sphere S^{k-1}

Let $f(x)$ $(x \in s^{k-1})$ be defined in some spherical neighborhood of a point $x^0 \in S^{k-1}$, i.e., on the set D^{k-1} $(x^0;\rho)$, and be integrable on the spheres $C^{k-2}(x^0;h)$ for all $h < \rho$. If there exist numbers $a_0, a_1, \dots, a_r$ such that the equality

$$\begin{aligned} &\frac{1}{|S^{k-2}|\sin^{k-2} h} \int_{C^{k-2}(x^0;h)} f(t)dS^{k-2}(t) \\ &= \frac{\Gamma\left(\frac{k-1}{2}\right)}{2\pi^{\frac{k-1}{2}}\sin^{k-2} h} \int_{C^{k-2}(x^0;h)} f(t)dS^{k-2}(t) \\ &= \sum_{v=0}^{r} \frac{\Gamma\left(\frac{k-1}{2}\right)a_v}{v!2^v\Gamma\left(\frac{k-1}{2}+v\right)}(1-\cosh)^v + o(1-\cosh)^r, \end{aligned} \tag{3.1}$$

is fulfilled in the neighborhood of a point x^0 ([109] and [110]), then we say that the function $f(x)$ has at the point x^0 a generalized Laplace operator of order r on the unit sphere S^{k-1} and denote it by $\overline{\Delta}^r f(x^0)$. The generalized Laplace operator defined by the equality (3.1) is related to the numbers a_v $(v = 0, r)$ as follows:

$$\left.\begin{array}{l}\overline{\Delta}^0 f(x^0) = a_0 = f(x^0),\\ \overline{\Delta}[\overline{\Delta}+1\cdot(k-1)][\overline{\Delta}+2\cdot k]\dots[\overline{\Delta}+(v-1)(v+k-3)]f(x^0) = a_v,\\ v=\overline{1,r},\end{array}\right\} \qquad (3.2)$$

where $\overline{\Delta}\cdot\overline{\Delta}\dots\overline{\Delta}$ (i-times) $= \overline{\Delta}^i$ and $\overline{\Delta}^i\cdot\overline{\Delta}^j = \overline{\Delta}^{i+j}$ for $i,j \geq 0$ and $i+j \leq r$.

Thus,by virtue of the decomposition (3.1), we have

$$\frac{1}{|S^{k-2}|\sin^{k-2}h}\int\limits_{C^{k-2}(x^0;h)} f(t)dS^{k-2}(t) = \overline{\Delta}^0 f(x^0)$$

$$+\sum_{v=1}^{r}\frac{\Gamma\left(\frac{k-1}{2}\right)\overline{\Delta}[\overline{\Delta}+1\cdot(k-1)][\overline{\Delta}+2\cdot k]\dots[\overline{\Delta}+(v-1)(v+k-3)]f(x^0)}{v!2^v\Gamma\left(\frac{k-1}{2}+v\right)}$$

$$\times(1-\cos h)^v + o(1-\cos h)^r, \quad r = 1,2,\dots.$$

Expansions of type (3.1) were considered for $k=3$ in [69] (p. 284).

It is easy to see that

$$a_1 = \overline{\Delta}^1 f(x^0) = \overline{\Delta} f(x^0)$$

$$= \lim_{h\to 0}\frac{\frac{1}{|S^{k-2}|\sin^{k-2}h}\int_{(x^0,t)=\cos h} f(t)dS^{k-2}(t) - f(x^0)}{\frac{2}{k-1}\sin^2\frac{h}{2}}.$$

Clearly, if there exists $\overline{\Delta}^r f(x)$, then $\overline{\Delta}^s f(x)$ exists for any $0 \leq s < r$.

Let us now introduce a more generalized Laplace operator on the sphere S^{k-1}.

We say that an integrable function $f(x)$ in the spherical neighborhood of a point $x^0 \in S^{k-1}$ has, at x^0, a more generalized Laplace operator $\widetilde{\Delta}^r f(x^0)$ of order r if the representation

$$\frac{1}{|D^{k-1}(x^0;h)|}\int\limits_{D^{k-1}(x^0;h)} f(t)dS^{k-1}(t)$$

$$= \sum_{v=0}^{r}\frac{\Gamma\left(\frac{k+1}{2}\right)b_v}{v!2^v\Gamma\left(\frac{k+1}{2}+v\right)}(1-\cos h)^v + o(1-\cos h)^r, \quad r = 0,1,\dots, \qquad (3.3)$$

holds.

A more generalized Laplace operator $\widetilde{\Delta}$ defined by the equality (3.3) is related to the coefficients b_v $(v=\overline{0,r})$ through (3.2).

Clearly, $b_0 = f(x^0)$ and

$$b_1 = \widetilde{\Delta}^1 f(x^0) = \widetilde{\Delta} f(x^0)$$

$$= \lim_{h\to 0} \frac{\frac{1}{|D^{k-1}(x^0;h)|}\int_{D^{k-1}(x^0;h)} f(t)dS^{k-1}(t) - f(x^0)}{\frac{2}{k+1}\sin^2\frac{h}{2}}.$$

Lemma 3.3.1. (a) *If at the point x^0 there exists $\overline{\Delta}^r f(x^0)$, then there also exists $\widetilde{\Delta}^r f(x^0)$, and $\overline{\Delta}^r f(x^0) = \widetilde{\Delta}^r f(x^0)$.*

(b) *There exists a function f for which $\widetilde{\Delta} f(x^0)$ is finite, but $\overline{\Delta} f(x)$ does not exist.*

Proof. Given

$$V(\rho) = \frac{1}{|S^{k-2}|\sin^{k-2}\rho}\int\limits_{C^{k-2}(x^0;\rho)} f(t)dS^{k-2}(t)$$

$$-\sum_{v=0}^{r}\frac{\Gamma\Big(\frac{k-1}{2}\Big)a_v}{v!2^v\Gamma\Big(\frac{k-1}{2}+v\Big)}(1-\cos\rho)^v = o(1-\cos\rho)^r, \tag{3.4}$$

we have to prove that

$$U(h) = \frac{1}{\int_0^h \sin^{k-2}\rho d\rho}\int\limits_0^h \sin^{k-2}\rho d\rho\Big\{\frac{1}{|S^{k-2}|\sin^{k-2}\rho}\int\limits_{C^{k-2}(x^0;\rho)} f(t)dS^{k-2}(t)\Big\}$$

$$-\sum_{v=0}^{r}\frac{\Gamma\Big(\frac{k+1}{2}\Big)a_v}{v!2^v\Gamma\Big(\frac{k+1}{2}+v\Big)}(1-\cosh)^v = o(1-\cosh)^r. \tag{3.5}$$

Since

$$\Gamma\Big(\frac{k+1}{2}\Big) = \frac{k-1}{2}\Gamma\Big(\frac{k-1}{2}\Big),$$

$$\Gamma\Big(\frac{k+1}{2}+v\Big) = \frac{2v+k-1}{2}\Gamma\Big(\frac{k-1}{2}+v\Big),$$

therefore

$$U(h) = \frac{1}{\int_0^h \sin^{k-2}\rho d\rho}\int\limits_0^h \sin^{k-2}\rho d\rho\Big\{\frac{1}{|S^{k-2}|\sin^{k-2}\rho}\int\limits_{C^{k-2}(x^0;\rho)} f(t)dS^{k-2}(t)\Big\}$$

$$-\sum_{v=0}^{r}\frac{\Gamma\Big(\frac{k-1}{2}\Big)a_v}{v!2^v\Gamma\Big(\frac{k-1}{2}+v\Big)}\cdot\frac{k-1}{2v+k-1}(1-\cosh)^v. \tag{3.6}$$

It is easy to verify that

$$\lim_{h\to 0}\frac{\int_0^h(1-cos\rho)^v\sin^{k-2}\rho d\rho}{(1-cosh)^v\int_0^h\sin^{k-2}\rho d\rho}=\frac{k-1}{2v+k-1}.$$

Thus

$$\frac{\int_0^h(1-\cos\rho)^v\sin^{k-2}\rho d\rho}{(1-\cosh)^v\int_0^h\sin^{k-2}\rho d\rho}=\frac{k-1}{2v+k-1}(1+\alpha_v(h)), \tag{3.7}$$

$\alpha_v(h)\to 0$, as $h\to 0$ for any v.

In view of (3.7), we have

$$(1-\cosh)^v=\frac{\int_0^h(1-\cos\rho)^v\sin^{k-2}\rho d\rho}{\frac{k-1}{2v+k-1}(1+\alpha_v(h))\int_0^h\sin^{k-2}\rho d\rho}.$$

Hence by virtue of (3.4), from (3.6) we get

$$\begin{aligned}U(h)=&\frac{1}{\int_0^h\sin^{k-2}\rho d\rho}\int_0^h\sin^{k-2}\rho d\rho\Big\{\frac{1}{|S^{k-2}|\sin^{k-2}\rho}\int_{C^{k-2}(x_0;\rho)}f(t)dS^{k-2}(t)\Big\}\\&-\sum_{v=0}^r\frac{\Gamma\Big(\frac{k-1}{2}\Big)a_v}{v!2^v\Gamma\Big(\frac{k-1}{2}+v\Big)}\cdot\frac{\int_0^h(1-cos\rho)^v\sin^{k-2}\rho d\rho}{(1+\alpha_v(h))\int_0^h\sin^{k-2}\rho d\rho}\\&=\frac{1}{\int_0^h\sin^{k-2}\rho d\rho}\int_0^h\Big\{\frac{1}{|S^{k-2}|\sin^{k-2}\rho}\int_{C^{k-2}(x^0;\rho)}f(t)dS^{k-2}(t)\\&-\sum_{v=0}^r\frac{\Gamma\Big(\frac{k-1}{2}\Big)\alpha_\nu}{v!2^v\Big(\frac{k-1}{2}+v\Big)(1+\alpha_v(h))}(1-\cos\rho)^v\Big\}\sin^{k-2}\rho d\rho\\&=\frac{1}{\int_0^h\sin^{k-2}\rho d\rho}\int_0^h o(1-\cos\rho)^r\sin^{k-2}\rho d\rho.\end{aligned}\tag{3.8}$$

By (3.7),

$$\begin{aligned}\int_0^h o(1-\cos\rho)^r\sin^{k-2}\rho d\rho&=o(1)\int_0^h(1-\cos\rho)^r\sin^{k-2}\rho d\rho\\&=o(1)(1-\cosh)^r\int_0^h\sin^{k-2}\rho d\rho.\end{aligned}$$

Hence it follows from (3.8) that

$$U(h) = \frac{1}{\int_0^h \sin^{k-2}\rho d\rho} \cdot o(1)(1-\cos h)^r \int\limits_0^h \sin^{k-2}\rho d\rho = o(1-\cos h)^r,$$

which proves the validity of Item (a), i.e., (3.5) is fulfilled.

(b) Define the function $f(x)$ as follows:

$$f(x) = \begin{cases} 0, & \text{when } \; x = x^0, \\ 1, & \text{when } \; (x^0, x) = \cos\gamma \; \text{ is rational}, \\ 0, & \text{when } \; (x^0, x) = \cos\gamma \; \text{ is irrational}. \end{cases}$$

It is easy to verify that the operator $\overline{\Delta} f(x^0)$ does not exist, and $\widetilde{\Delta} f(x^0) = 0$. The lemma is proved. □

Let us introduce the notion of a strong generalized Laplace operator of order r on S^{k-1}. Let the function $f(x)$, $x \in S^{k-1}$ be defined in some spherical neighborhood of the point x^0, i.e., on the set $D^{k-1}(x^0;\rho)$, and be integrable on the spheres $C^{k-2}(x^0;h)$ for all $h < \rho$. If there exist functions $a_i(x)$ $i = \overline{0, r-1}$ and a number a_r such that there exist finite limits $\lim\limits_{x\to x^0} a_i(x) = a_i$, and in the neighborhood of the point x^0 the inequality

$$\begin{aligned} \frac{1}{|S^{k-2}|\sin^{k-2}h} \int\limits_{C^{k-2}(x;h)} f(t)dS^{k-2}(t) &= \sum_{v=0}^{r-1} \frac{\Gamma\left(\frac{k-1}{2}\right)a_v(x)}{v!2^v\Gamma\left(\frac{k-1}{2}+v\right)}(1-\cos h)^v \\ &+ \frac{\Gamma\left(\frac{k-1}{2}\right)a_r}{r!2^r\Gamma\left(\frac{k-1}{2}+r\right)}(1-\cos h)^r + \varepsilon(h,x)(1-\cos h)^r, \end{aligned} \tag{3.9}$$

holds, where $\lim\limits_{\substack{h\to 0\\ x\to x^0}} \varepsilon(h,x) = 0$, then we say that the function $f(x)$ at the point $x^0 \in S^{k-1}$ has a strong generalized Laplace operator of order r, which we denote by the symbol $\overline{\Delta}_x^r f(x^0)$. Clearly, if $x = x^0$, then $\overline{\Delta}_x^r f(x^0) = \overline{\Delta}^r f(x^0)$.

The strong generalized Laplace operator $\overline{\Delta}_x^r$ defined by the equality (3.9) is related to the numbers a_v $(v = \overline{0,r})$ through (3.2).

It is obvious that

$$\overline{\Delta}_x f(x^0) = \lim_{(x,h)\to(x^0,0)} \frac{\frac{1}{|S^{k-2}|\sin^{k-2}h}\int_{C^{k-2}(x;h)} f(t)dS^{k-2}(t) - f(x)}{\frac{2}{k-1}\sin^2\frac{h}{2}}.$$

Let us now introduce a more generalized strong Laplace operator on S^{k-1}. Let in the spherical neighborhood of the point $x^0 \in S^{k-1}$ there exist functions $b_i(x)$

$i = \overline{0, r-1}$ and a number b_r such that there exist limits $\lim\limits_{x\to x_0} b_i(x) = b_i$. We say that an integrable function $f(x)$ in the neighborhood of a point $x^0 \in S^{k-1}$ has, at x^0, a more generalized strong Laplace operator $\widetilde{\Delta}_x^r f(x^0)$ of order r if the equality

$$\frac{1}{|D^{k-1}(x^0;h)|} \int\limits_{D^{k-1}(x;h)} f(t)dS^{k-1}(t) = \sum_{v=0}^{r-1} \frac{\Gamma\left(\frac{k+1}{2}\right)b_v(x)}{v!2^v\Gamma\left(\frac{k+1}{2}+v\right)}(1-\cos h)^v$$
$$+\frac{\Gamma\left(\frac{k+1}{2}\right)b_r}{r!2^r\Gamma\left(\frac{k+1}{2}+r\right)}(1-\cos h)^r + \varepsilon(h,x)(1-\cos h)^r, \tag{3.10}$$

holds, where $\lim\limits_{\substack{h\to 0\\ x\to x^0}} \varepsilon(h,x) = 0$. Clearly, if $x = x^0$, then $\widetilde{\Delta}_x^r f(x^0) = \widetilde{\Delta}^r f(x^0)$.

A more generalized strong Laplace operator $\widetilde{\Delta}_x^r$ defined by the equality (3.10) is related to b_v $(v = \overline{0, r})$ through (3.2).

It is not difficult to see that

$$\widetilde{\Delta}_x f(x^0) = \lim_{(x,h)\to(x^0,0)} \frac{\frac{1}{|D^{k-1}(x_0;h)|}\int_{D^{k-1}(x;h)} f(t)dS^{k-1}(t) - f(x)}{\frac{2}{k+1}\sin^2\frac{h}{2}}.$$

The notion of a generalized Laplace operator on the unit sphere S^{k-1} is widely used in problems of the summability of differentiated Fourier–Laplace series. A Laplace operator on the unit sphere or rather the angular part of the Laplace operator written in terms of spherical coordinates is called a differentiation operator.

Lemma 3.3.2. *For Gegenbauer polynomials $P_n^\lambda(x)$, the representation*

$$P_n^\lambda(\cos h) = P_n^\lambda(1)\Bigg\{1$$
$$+\sum_{v=1}^{\infty} \frac{[-n(n+2\lambda)][-n(n+2\lambda)+1\cdot(k-1)]\dots[-n(n+2\lambda)+(v-1)(k+v-3)]}{v!2^v}$$
$$\times\frac{\Gamma\left(\frac{k-1}{2}\right)}{\Gamma\left(\frac{k-1}{2}+\nu\right)}(1-\cos h)^v\Bigg\} \tag{3.11}$$

is valid.

Proof. For Gegenbouer polynomials $P_n^\lambda(x)$ we have the representation (see [58], p. 92): $P_n^\lambda(x) = P_n^\lambda F(-n, n+2\lambda; \lambda+\frac{1}{2}; \frac{1-x}{2})$, where $F(a,b;c;x)$ is the hypergeometric series (see [58], p. 74)

$$F(a,b;c;x) = 1 + \sum_{v=1}^{\infty} \frac{a(a+1)\dots(a+v-1)}{v!}\cdot\frac{b(b+1)\dots(b+v-1)}{c(c+1)\dots(c+v-1)}c^v.$$

Therefore

$$P_n^\lambda(x) = P_n^\lambda(1)\Big[1 + \frac{-n}{!}\cdot\frac{n+2\lambda}{\lambda+\frac{1}{2}}\cdot\frac{1-x}{2} + \frac{-n(-n+1)}{2!}$$
$$\times\frac{(n+2\lambda)(n+2\lambda+1)}{(\lambda+\frac{1}{2})(\lambda+\frac{3}{2})}\cdot\frac{(1-x)^2}{2^2} + \frac{-n(-n+1)(-n+2)}{3!}$$
$$\times\frac{(n+2\lambda)(n+2\lambda+1)(n+2\lambda+2)}{(\lambda+\frac{1}{2})(\lambda+\frac{3}{2})(\lambda+\frac{5}{2})}\cdot\frac{(1-x)^3}{2^3} + \cdots$$
$$+\frac{-n(-n+1)(-n+2)\ldots(-n+r-1)}{r!}$$
$$\times\frac{(n+2\lambda)(n+2\lambda+1)\ldots(n+2\lambda+r-1)}{(\lambda+\frac{1}{2})(\lambda+\frac{3}{2})\ldots(\lambda+r-\frac{1}{2})}\cdot\frac{(1-x)^r}{2^r} + \cdots\Big].$$

But

$$-n(-n+1)(n+2\lambda)(n+2\lambda+1) = -n(n+2\lambda)[-n(n+2\lambda)+1\cdot(k-1)],$$
$$-n(-n+1)(-n+2)(n+2\lambda)(n+2\lambda+1)(n+2\lambda+2)$$
$$= -n(n+2\lambda)[-n(n+2\lambda)+1\cdot(k-1)][-n(n+2\lambda)+2\cdot k]$$

. . . .

$$-n(-n+1)\ldots(-n+r-1)(n+2\lambda)(n+2\lambda+1)\ldots(n+2\lambda+r-1)$$
$$= -n(n+2\lambda)[-n(n+2\lambda)+1\cdot(k-1)][-n(n+2\lambda)+2\cdot k]\times\cdots$$
$$\times[-n(n+2\lambda)+(r-1)(r+k-3)].$$

Moreover,

$$\lambda+\frac{1}{2} = \frac{k-1}{2} = \frac{\Gamma\Big(\frac{k-1}{2}+1\Big)}{\Gamma\Big(\frac{k-1}{2}\Big)},$$

$$\Big(\lambda+\frac{1}{2}\Big)\Big(\lambda+\frac{3}{2}\Big) = \frac{k-1}{2}\cdot\frac{k+1}{2} = \frac{\Gamma\Big(\frac{k-1}{2}+2\Big)}{\Gamma\Big(\frac{k-1}{2}\Big)},$$

$$\Big(\lambda+\frac{1}{2}\Big)\Big(\lambda+\frac{3}{2}\Big)\Big(\lambda+\frac{5}{2}\Big) = \frac{k-1}{2}\cdot\frac{k+1}{2}\frac{k+3}{2} = \frac{\Gamma\Big(\frac{k-1}{2}+3\Big)}{\Gamma\Big(\frac{k-1}{2}\Big)},$$

. . . .

$$\Big(\lambda+\frac{1}{2}\Big)\Big(\lambda+\frac{3}{2}\Big)\cdot\Big(\lambda+r-\frac{1}{2}\Big) = \frac{k-1}{2}\cdot\frac{k+1}{2}\cdots\frac{k+2r-3}{2} = \frac{\Gamma\Big(\frac{k-1}{2}+r\Big)}{\Gamma\Big(\frac{k-1}{2}\Big)}.$$

Taking these equalities into account, we obtain the equalities (3.11). Lemma 3.3.2 is proved. □

Lemma 3.3.3. *If $Y_n(x)$ is a spherical harmonic of order n, then*

$$\frac{1}{P_n^\lambda(1)}P_n^\lambda(\cosh)Y_n(x)=\frac{1}{|C^{k-2}(x;h)|}\int\limits_{C^{k-2}(x;h)}Y_n(t)dS^{k-2}(t). \tag{3.12}$$

Proof. As is known ([66], p. 163),

$$Y_n(t)=\frac{(n+\lambda)\Gamma(\lambda)}{2\pi^{\lambda+1}}\int\limits_{S^{k-1}}P_n^\lambda[(t,\eta)]Y_n(\eta)dS^{k-1}(\eta),$$

whence

$$\int\limits_{C^{k-2}(x,h)}Y_n(t)dS^{k-2}(t)$$

$$=\frac{(n+\lambda)\Gamma(\lambda)}{2\pi^{\lambda+1}}\int\limits_{C^{k-2}(x,h)}dS^{k-2}(t)\int\limits_{S^{k-1}}Y_n(\eta)P_n^\lambda[(t,\eta)]dS^{k-1}(\eta)$$

$$=\frac{(n+\lambda)\Gamma(\lambda)}{2\pi^{\lambda+1}}\int\limits_{S^{k-1}}Y_n(\eta)dS^{k-1}(\eta)\int\limits_{C^{k-2}(x;h)}P_n^\lambda[(t,\eta)]dS^{k-2}(t). \tag{3.13}$$

Using the addition theorem ([6], p. 467), from (3.3) we obtain

$$\int\limits_{C^{k-2}(x;h)}Y_n(t)dS^{k-2}(t)=\frac{n!\Gamma(2\lambda)}{\Gamma(n+2\lambda)}$$

$$\times\frac{(n+\lambda)\Gamma(\lambda)}{2\pi^{\lambda+1}}|C^{k-2}(x;h)|P_n^\lambda[(t,x)]\int\limits_{S^{k-1}}Y_n(\eta)P_n^\lambda[(x,\eta)]dS^{k-1}(\eta)$$

$$=\frac{1}{P_n^\lambda(1)}|C^{k-2}(x;h)|P_n^\lambda[(t,x)]Y_n(x),$$

whence

$$\frac{1}{P_n^\lambda(1)}P_n^\lambda(\cosh)Y_n(x)=\frac{1}{|C^{k-2}(x;h)|}\int\limits_{C^{k-2}(x,y)}Y_n(t)dS^{k-2}(t).$$

Lemma 3.3.3 is proved. □

These lemmas underlie the proof of

Theorem 3.3.1. *If $Y_n(x)$, $n=0,1,2,\dots$ is an arbitrary spherical harmonic of order n, and ξ is an arbitrary point on S^{k-1}, then for each nonnegative integer r there exists $\overline{\Delta}^rY_n(\xi)$, and the equality $\overline{\Delta}^rY_n(\xi)=D_k^rY_n(\xi)$ holds.*

Proof. Since $\overline{\Delta}^r$ and D_k^r are linear operators, it suffices to show that $\overline{\Delta}^r Y_n(\xi)$ exists and $\overline{\Delta}^r Y_n(\xi) = D_k^r Y_n(\xi)$, where $Y_n(x)$ is normalized so that $Y_n(\xi) = P_n^\lambda(1) = \binom{n+2\lambda-1}{n}$. Then $D_k^r Y_n(\xi) = D_k^r P_n^\lambda[(\xi, x)]_{x=\xi}$ (because both operators are equal to $[-n(n+2\lambda)]^r P_n^\lambda(1)$).

By the normalization of $Y_n(x)(Y_n(\xi) = P_n^\lambda(1))$, from (3.12) we obtain

$$P_n^\lambda(\cosh) = \frac{1}{|C^{k-2}(\xi;h)|} \int\limits_{C^{k-2}(\xi;h)} Y_n(t) dS^{k-2}(t). \tag{3.14}$$

On the other hand,

$$P_n^\lambda(\cosh) = P_n^\lambda[(\xi,t)] = \frac{1}{|C^{k-2}(\xi;h)|} \int\limits_{C^{k-2}(\xi;h)} P_n^\lambda[(\xi,t)] dS^{k-2}(t). \tag{3.15}$$

From (3.14) and (3.15), we have

$$\begin{aligned} \frac{1}{|C^{k-2}(\xi;h)|} \int\limits_{C^{k-2}(\xi;h)} Y_n(t) dS^{k-2}(t) &= P_n^\lambda(\cos h) \\ &= \frac{1}{|C^{k-2}(\xi;h)|} \int\limits_{C^{k-2}(\xi;h)} P_n^\lambda[(\xi,t)] dS^{k-2}(t), \end{aligned}$$

or, which is the same,

$$\overline{\Delta}^r Y_n(\xi) = \overline{\Delta}^r P_n^\lambda[(\xi,t)]_{t=\xi}.$$

Thus to prove the theorem, it suffices to prove the equality

$$\overline{\Delta}^r P_n^\lambda[(\xi,t)]_{t=\xi} = [-n(n+2\lambda)]^r P_n^\lambda(1).$$

But this follows from the fact that the equality (3.11) holds,

$$\begin{aligned} &\frac{1}{|C^{k-2}(\xi;h)|} \int\limits_{C^{k-2}(\xi;h)} P_n^\lambda[(\xi,t)] dS^{k-2}(t) = P_n^\lambda(\cosh) = P_n^\lambda(1)\Bigg\{1 \\ &+ \sum_{v=1}^{\infty} \frac{[-n(n+2\lambda)][-n(n+2\lambda)+1\cdot(k-1)]\ldots[-n(n+2\lambda)+(v-1)(k+v-3)]}{v!2^v} \\ &\times \frac{\Gamma\left(\frac{k-1}{2}\right)}{\Gamma\left(\frac{k-1}{2}+v\right)}(1-\cosh)^v\Bigg\}. \end{aligned}$$

Thus we finally have

$$D_k^r Y_n(\xi) = D_k^r P_n^\lambda[(\xi,t)]_{t=\xi} = \overline{\Delta}^r P_n^\lambda[(\xi,t)]_{t=\xi} = \overline{\Delta}^r Y_n(\xi).$$

Therefore $\overline{\Delta}^r Y_n(\xi) = D_k^r Y_n(\xi)$.

Theorem 3. 3.1 is proved. □

Theorem 3.3.2. *If a function $f(x)$, $x \in S^{k-1}$ has continuous partial derivatives of order $2r$ in the neighborhood of a point $x^0 \in S^{k-1}$, then $\overline{\Delta}^r f(x^0)$ exists and $\overline{\Delta}^r f(x^0) = D_k^r f(x^0)$.*

Theorem 3.3.2 is proved as an analogous theorem for $k = 3$ proved in [69] (p. 306).

3.4 Boundary Properties of the Integral $D_k U(f; \rho, \theta_1, \theta_2, \dots, \theta_{k-2}, \varphi)$

The section will deal with the boundary properties of the integral $D_k(f; \rho, \theta_1, \theta_2, \dots, \theta_{k-2}, \varphi)$, where D_k is the Laplace operator on the sphere S^{k-1}, i.e., the angular part of the Laplace operator written in terms of spherical coordinates ([30, 77–86],[95],[100],[106]).

We can show that the function $P(\rho, \gamma) = \sum_{n=0}^{\infty} \frac{n+\lambda}{\lambda} P_n^{\lambda}(\cos\gamma)\rho^n$ is harmonic in V^k.

Indeed, taking (1.10) into account, we have

$$\begin{aligned}
\Delta P(\rho,\gamma) &= \frac{1}{\rho^2} D_k \Big(\sum_{n=0}^{\infty} \frac{n+\lambda}{\lambda} P_n^{\lambda}(\cos\gamma)\rho^n \Big) \\
&+ \frac{1}{\rho^{k-1}} \frac{\partial}{\partial \rho} \Big[\rho^{k-1} \frac{\partial}{\partial \rho} \Big(\sum_{n=0}^{\infty} \frac{n+\lambda}{\lambda} P_n^{\lambda}(\cos\gamma)\rho^n \Big) \Big] \\
&= -\frac{1}{\rho^2} \Big[\sum_{n=1}^{\infty} \frac{n+\lambda}{\lambda} n(n+k-2) P_n^{\lambda}(\cos\gamma)\rho^n \Big] \\
&+ \frac{1}{\rho^{k-1}} \frac{\partial}{\partial \rho} \Big[\rho^{k-1} \Big(\sum_{n=1}^{\infty} \frac{n+\lambda}{\lambda} n P_n^{\lambda}(\cos\gamma)\rho^{n-1} \Big) \Big] \\
&= -\frac{1}{\rho^2} \Big[\sum_{n=1}^{\infty} \frac{n+\lambda}{\lambda} n(n+k-2) P_n^{\lambda}(\cos\gamma)\rho^n \Big] \\
&+ \frac{1}{\rho^{k-1}} \Big[\sum_{n=1}^{\infty} \frac{n+\lambda}{\lambda} n(n+k-2) P_n^{\lambda}(\cos\gamma)\rho^{n+k-3} \Big] \\
&= -\frac{1}{\rho^2} \Big[\sum_{n=1}^{\infty} \frac{n+\lambda}{\lambda} n(n+k-2) P_n^{\lambda}(\cos\gamma)\rho^n \Big] \\
&+ \frac{1}{\rho^2} \Big[\sum_{n=1}^{\infty} \frac{n+\lambda}{\lambda} n(n+k-2) P_n^{\lambda}(\cos\gamma)\rho^n \Big] = 0.
\end{aligned}$$

Therefore (see (1.5))

$$D_k P(\rho,\gamma) = \frac{1}{\rho^{k-3}} \cdot \frac{\partial}{\partial \rho} \Big[\rho^{k-1} \frac{\partial P(\rho,\gamma)}{\partial \rho} \Big]. \tag{4.1}$$

Lemma 3.4.1. *If $P(\rho,\gamma)$ is the Poisson kernel, then*

$$D_kP(\rho,\gamma)=\frac{A(\rho,\gamma)}{[(1-\rho)^2+4\sin^2\frac{\gamma}{2}]^{\lambda+3}}, \tag{4.2}$$

where all terms in $A(\rho,\gamma)$ with respect to $(1-\rho)$ and $\sin\frac{\gamma}{2}$ are of degree ≥ 3, and $a(\rho,\gamma)$ is divided by $(1-\rho^2)$.

Proof. Let

$$\rho(\rho,\gamma)=(1-\rho^2)[\Delta(\rho,\gamma)]^{-(\lambda+1)}, \tag{4.3}$$

where

$$\Delta(\rho,\gamma)=(1-\rho)^2+4\rho\sin^2\frac{\gamma}{2}.$$

By (4.1) and (4.3), we have

$$\begin{aligned}D_kP(\rho,\gamma)&=-\frac{1}{\rho^{k-3}}\cdot\frac{\partial}{\partial\rho}\Big[\rho^{k-1}\frac{\partial P(\rho,\gamma)}{\partial\rho}\Big]\\&=-\frac{1}{\rho^{k-3}}\frac{\partial}{\partial\rho}\rho^{k-1}\Big\{-2\rho[\Delta(\rho,\gamma)]^{-(\lambda+1)}\\&-(\lambda+1)(1-\rho^2)[\Delta(\rho,\gamma)]^{-(\lambda+2)}\frac{\partial\Delta(\rho,\gamma)}{\partial\rho}\Big\}\\&=\frac{1}{\rho^{k-3}}\frac{\partial}{\partial\rho}\{B(\rho,\gamma)[\Delta(\rho,\gamma)]^{-(\lambda+2)}\},\end{aligned} \tag{4.4}$$

where

$$B(\rho,\gamma)=\rho^{k-1}[2\rho\Delta(\rho,\gamma)+(\lambda+1)(1-\rho^2)\frac{\partial\Delta(\rho,\gamma)}{\partial\rho}].$$

It is obvious that all terms in $B(\rho,\gamma)$ with respect to $(1-\rho)$ and $\sin\frac{\gamma}{2}$ are of degree ≥ 2.

From (4.4) we have

$$\begin{aligned}D_kP(\rho,\gamma)&=\frac{1}{\rho^{k-3}}\Big\{\frac{\partial B(\rho,\gamma)}{\partial\rho}[\Delta(\rho,\gamma)^{-(\lambda+2)}]\\&-(\lambda+2)[\Delta(\rho,\gamma)]^{-(\lambda+3)}\frac{\partial\Delta(\rho,\gamma)}{\partial\rho}\cdot B(\rho,\gamma)\Big\}\\&=A(\rho,\gamma)[\Delta(\rho,\gamma)]^{-(\lambda+3)}=(1-\rho^2)D_k[\Delta(\rho,\gamma)]^{-(\lambda+1)},\end{aligned} \tag{4.5}$$

where

$$A(\rho,\gamma)=\frac{1}{\rho^{k-3}}\Big[\frac{\partial B(\rho,\gamma)}{\partial\rho}\cdot\Delta(\rho,\gamma)-(\lambda+2)B(\rho,\gamma)\frac{\partial\Delta(\rho,\gamma)}{\partial\rho}\Big].$$

It is not difficult to see that all terms in $A(\rho,\gamma)$ with respect to $(1-\rho)$ and $\frac{\gamma}{2}$ are of degree ≥ 3 and, as follows from (4.5), are divided by $(1-\rho^2)$.

Lemma 3.4.1 is proved. □

The property of $A(\rho,\gamma)$ implies

$$|A(\rho,\gamma)| \leq (1-\rho)I(\rho,\gamma), \tag{4.6}$$

where $I(\rho,\gamma)$ is a homogeneous polynomial of degree 2 of $(1-\rho,\gamma)$ with a positive coefficient.

Lemma 3.4.2. *The following statements are valid:*

1) $\int\limits_0^x D_k P(\rho,\gamma)\sin^{k-2}\gamma d\gamma = 0;$

2) $\lim\limits_{\rho\to 1}\max\limits_{0<\delta\leq\gamma\leq\pi}|D_k P(\rho,\gamma)| = 0;$

3) $\int\limits_0^\pi D_k P(\rho,\gamma)\sin^{k-2}\gamma\sin^2\frac{\gamma}{2}d\gamma = \dfrac{\rho(k-1)\pi^{\frac{1}{2}}\Gamma\left(\frac{k-1}{2}\right)}{2\Gamma\left(\frac{k}{2}\right)};$

4) $\int\limits_0^\pi \gamma^k|D_k P(\rho,\gamma)|d\gamma = O(1).$

Proof. Statements 1) and 2) are obvious. Let us prove the property 3). By (4.1), from (2.6) we have (4.7)

$$\begin{aligned} D_k P(\rho,\gamma) &= -(k-1)\rho\frac{\partial P(\rho,\gamma)}{\partial\rho} - \rho^2\frac{\partial^2 P(\rho,\gamma)}{\partial\rho^2} \\ &= -\frac{k-1}{\lambda}\sum_{n=1}^{\infty} n(n+\lambda)P_n^\lambda(\cos\gamma)\rho^n \\ &\quad -\frac{1}{\lambda}\sum_{n=2}^{\infty}(n-1)n(n+\lambda)P_n^\lambda(\cos\gamma)\rho^n. \end{aligned} \tag{4.7}$$

Taking into account the equality

$$\sin^2\frac{\gamma}{2} = \frac{1-\cos\gamma}{2} = \frac{2\lambda - P_1^\gamma(\cos\gamma)}{4\lambda},$$

from (4.7) we obtain

$$\begin{aligned} \int\limits_{S^{k-1}} D_k P(\rho,\gamma)\sin^2\frac{\gamma}{2}dS^{k-1}(y) &= \frac{\rho(k-1)(\lambda+1)}{4\lambda^2}\int\limits_{S^{k-1}}[P_1^\lambda(\cos\gamma)]^2 dS^{k-1}(y) \\ &= \rho(k-1)(\lambda+1)\int\limits_{S^{k-1}}\cos^2\gamma dS^{k-1}(y) \\ &= \rho(k-1)(\lambda+1)\int\limits_0^\pi\cos^2\gamma d\gamma\int\limits_{(x,y)=\cos\gamma} dS^{k-2}(y) \end{aligned}$$

$$= \rho(k-1)(\lambda+1)\int_0^{\pi}\frac{2\pi^{\frac{k-1}{2}}\sin^{k-2}\gamma}{\Gamma\left(\frac{k-1}{2}\right)}\cos^2\gamma d\gamma$$

$$= \frac{2\rho(k-1)(\lambda+1)\pi^{\frac{k-1}{2}}}{\Gamma\left(\frac{k-1}{2}\right)}\int_0^{\pi}(\sin^{k-2}\gamma - \sin^k\gamma)d\gamma$$

$$= \frac{2\rho(k-1)(\lambda+1)\pi^{\frac{k-1}{2}}}{\Gamma\left(\frac{k-1}{2}\right)}\left[\frac{\pi^{\frac{1}{2}}\Gamma\left(\frac{k-1}{2}\right)}{\Gamma\left(\frac{k}{2}\right)} - \frac{\pi^{\frac{1}{2}}\Gamma\left(\frac{k+1}{2}\right)}{\Gamma\left(\frac{k+2}{2}\right)}\right]$$

$$= \frac{2\rho(k-1)(\lambda+1)\pi^{\frac{k}{2}}}{\Gamma\left(\frac{k-1}{2}\right)}\left[\frac{\Gamma\left(\frac{k-1}{2}\right)}{\Gamma\left(\frac{k}{2}\right)} - \frac{\frac{k-1}{2}\Gamma\left(\frac{k-1}{2}\right)}{\frac{k}{2}\Gamma\left(\frac{k+2}{2}\right)}\right] = \frac{2\rho(k-1)(\lambda+1)\pi^{\frac{k}{2}}}{k\Gamma\left(\frac{k}{2}\right)}.$$

Consequently,

$$\int_0^{\pi} D_kP(\rho,\gamma)\sin^2\frac{\gamma}{2}d\gamma \int_{(x,y)=\cos\gamma} dS^{k-2}(y) = \frac{2\rho(k-1)(\lambda+1)\pi^{\frac{k}{2}}}{k\Gamma\left(\frac{k}{2}\right)},$$

whence

$$\frac{2\pi^{\frac{k-1}{2}}}{\Gamma\left(\frac{k-1}{2}\right)}\int_0^{\pi} D_kP(\rho,\gamma)]\sin^{k-2}\gamma\sin^2\frac{\gamma}{2}d\gamma = \frac{2\rho(k-1)(\lambda+1)\pi^{\frac{k}{2}}}{k\Gamma\left(\frac{k}{2}\right)}. \tag{4.8}$$

Statement 3) follows from (4.8).

Using Lemma 3.4.1, we can now prove the validity of statement 4). Taking into account (4.6) and the inequality $|\sin\gamma| \geq \frac{2}{\pi}|\gamma|$, for $|\gamma| \leq \frac{\pi}{2}$, we obtain

$$\int_0^{\pi}\gamma^k|D_kP(\rho,\gamma)|d\gamma \leq (1-\rho)\int_0^{\pi}\frac{I(\rho,\gamma)\gamma^k d\gamma}{\left[(1-\rho)^2 + 4\rho\sin^2\frac{\gamma}{2}\right]^{\lambda+3}}$$

$$\leq C(1-\rho)\int_0^{\pi}\frac{I(\rho,\gamma)\gamma^k d\gamma}{\left[\pi^2(1-\rho)^2 + 4\rho\gamma^2\right]^{\lambda+3}}. \tag{4.9}$$

Using the substitution $\gamma = (1-\rho)t$ and assuming $\rho > \frac{1}{2}$, from (4.9) we find

$$\int_0^{\pi}|D_k(\rho,\gamma)|\gamma^k d\gamma < C(1-\rho)\int_0^{\frac{\pi}{1-\rho}}\frac{(1-\rho)^2(1-\rho)^k(1-\rho)\left(\sum_{v=0}^2 C_v t^v\right)t^k dt}{(1-\rho)^{2\lambda+6}(1+t^2)^{\lambda+3}}$$

$$< C\int_0^{\infty}\frac{\left(\sum_{v=0}^2 t^{k+v}\right)dt}{(1+t^2)^{\lambda+3}} = O(1).$$

Statement 4) and Lemma 3.4.2 are proved. □

Theorem 3.4.1. (a) *If at the point $x^0(1,\theta_1^0,\theta_2^0,\dots,\theta_{k-2}^0,\varphi^0)$ there exists a finite $\overline{\Delta}f(x^0)$ (see (3.1)), then*

$$\lim_{\rho\to 1} D_k U(f;\rho,x^0)=\overline{\Delta}f(x^0).$$

(b) *There exists a function $f(\theta,\varphi)$ such that $D_3f(0,0)=0$, but the limit*

$$\lim_{(\rho,\theta,\varphi)\overset{\wedge}{\longrightarrow}(1,0,0)} D_3U(f;\rho,\theta,\varphi)$$

does not exist.

Proof of Item (a). Indeed, taking x^0 as an initial point, by Lemma 3.4.2 we have

$$\frac{1}{\rho}D_kU(f;\rho,x^0)=\frac{\Gamma\left(\frac{k}{2}\right)}{2\rho\pi^{\frac{k}{2}}}\int\limits_{S^{k-1}} D_kP(\rho,\gamma)f(y)dS^{k-1}(y)$$

$$=\frac{\Gamma\left(\frac{k}{2}\right)}{2\rho\pi^{\frac{k}{2}}}\int\limits_{S^{k-1}} D_kP(\rho,\gamma)[f(y)-f(x^0)]dS^{k-1}(y)$$

$$=\frac{\Gamma\left(\frac{k}{2}\right)}{2\rho\pi^{\frac{k}{2}}}\int\limits_0^{\pi} D_kP(\rho,\gamma)d\gamma\int\limits_{(x^0,y)=\cos\gamma}[f(y)-f(x^0)]dS^{k-2}(y)$$

$$=\frac{\Gamma\left(\frac{k}{2}\right)}{\rho\pi^{\frac{k}{2}}\Gamma\left(\frac{k-1}{2}\right)}\int\limits_0^{\pi}\left\{D_kP(\rho,\gamma)\sin^{2\lambda}\lambda\frac{1}{|S^{k-2}|\sin^{2\lambda}\gamma}\right.$$

$$\left.\times\int\limits_{(x^0,y)=\cos\gamma}[f(y)-f(x^0)]dS^{k-2}(y)\right\}d\gamma$$

$$=\frac{\Gamma\left(\frac{k}{2}\right)}{\rho\pi^{\frac{1}{2}}\Gamma\left(\frac{k-1}{2}\right)}\int\limits_0^{\pi} D_kP(\rho,\gamma)\sin^{2\lambda}\gamma\sin^2\frac{\gamma}{2}$$

$$\times\left\{\frac{2}{k-1}\cdot\frac{\frac{1}{|S^{k-2}|\sin^{2\lambda}\gamma}\int_{(x^0,y)=\cos\gamma}[f(y)-f(x^0)]dS^{k-2}(y)}{\frac{2}{k-1}\sin^2\frac{\gamma}{2}}\right\}d\gamma$$

$$=\frac{2\Gamma\left(\frac{k}{2}\right)}{\rho(k-1)\pi^{\frac{1}{2}}\Gamma\left(\frac{k-1}{2}\right)}\int\limits_0^{\pi} D_kP(\rho,\gamma)\sin^{2\lambda}\gamma\sin^2\frac{\lambda}{2}$$

$$\times\left\{\frac{\frac{1}{|S^{k-2}|\sin^{2\lambda}\gamma}\int\limits_{(x^0,y)=\cos\gamma}[f(y)-f(x^0)]dS^{k-2}(y)}{\frac{2}{k-1}\sin^2\frac{\gamma}{2}}-\overline{\Delta}f(x^0)\right\}d\gamma+\overline{\Delta}f(x^0)$$

Let $\varepsilon > 0$ be an arbitrary number. We choose $\delta > 0$ such that

$$|\Delta_\gamma(f;x^0)-\overline{\Delta}f(x^0)|<\varepsilon \quad \text{when} \quad 0<\gamma<\delta, \tag{4.10}$$

where

$$\Delta_r(f;x^0)=\frac{\frac{1}{|S^{k-2}|\sin^{2\lambda}\gamma}\int_{(x^0,y)=\cos\gamma}[f(y)-f(x^0)]dS^{k-2}(y)}{\frac{2}{k-1}\sin^2\frac{\gamma}{2}}. \tag{4.11}$$

Then

$$\frac{1}{\rho}D_kU(f;\rho,x^0)=I_1+I_2+\overline{\Delta}f(x^0).$$

Here

$$I_1=\frac{2\Gamma\left(\frac{k}{2}\right)}{\rho(k-1)\pi^{\frac{1}{2}}\Gamma\left(\frac{k-1}{2}\right)}\int\limits_0^\delta D_kP(\rho,\gamma)\sin^{2\lambda}\gamma\sin^2\frac{\gamma}{2}\,[\Delta_\gamma(f;x^0)-\overline{\Delta}f(x^0)]d\gamma,$$

$$I_2=\frac{2\Gamma\left(\frac{k}{2}\right)}{\rho(k-1)\pi^{\frac{k}{2}}\Gamma\left(\frac{k-1}{2}\right)}\int\limits_\delta^\pi D_kP(\rho,\gamma)\sin^{2\lambda}\gamma\sin^2\frac{\gamma}{2}\,[\Delta_\gamma(f;x^0)-\overline{\Delta}f(x^0)]d\gamma.$$

By Lemma 3.4.2 and the inequality (4.10), we have

$$|I_1|<C\varepsilon, \quad \text{when} \quad 0<\gamma\le\delta. \tag{4.12}$$

Further, by Lemma 3.4.2 it can be easily shown that

$$\lim_{\rho\to1}I_2=0. \tag{4.13}$$

(4.12) and (4.13) imply that item (a) of Theorem 3.4.1 is valid.

Proof of Item (b). It can be easily verified that in the space R^3,

$$D_3U(f;\rho,\theta,\varphi)$$

$$=\frac{1}{4\pi}\int\limits_0^\pi\int\limits_0^{2\pi}\frac{3\rho(1-\rho^2)[5\rho-\rho\cos^2\gamma-2(1+\rho^2)\cos\gamma]}{(1-2\rho\cos\gamma+\rho^2)^{\frac{7}{2}}}\cdot f(\theta',\varphi')\sin\theta'd\theta'd\varphi'$$

where

$$\cos\gamma=\cos\theta\cos\theta'+\sin\theta\sin\theta'\cos(\varphi-\varphi').$$

By Lemma 3.4.2, for any point $(1,\theta,\varphi)$,

$$\int\limits_0^\pi\int\limits_0^{2\pi}\frac{5\rho-\rho\cos^2\gamma-2(1+\rho^2)\cos\gamma}{(1-2\rho\cos\gamma+\rho^2)^{\frac{7}{2}}}\sin\theta'd\theta'd\varphi'=0. \tag{4.14}$$

Let now $\overline{x}(\rho,\theta,\varphi)\to(1,0,0)$ so that $\varphi=0$ and $\theta=1-\rho$. In this case,

$$\cos\gamma=\cos\theta\cos\theta'+\sin\theta\sin\theta'\cos\varphi',$$
$$A(\rho,\gamma)=5\rho-\rho\cos^2\gamma-2(1+\rho^2)\cos\gamma$$
$$=5\rho-2\rho\sin\theta\sin\theta'\cos\theta\cos\theta'\cos\varphi'-2(1+\rho^2)\sin\theta\sin\theta'\cos\varphi'$$
$$-[\rho\cos^2\theta\cos^2\theta'+\rho\sin^2\theta\sin^2\theta'\cos^2\varphi'+2(1+\rho^2)\cos\theta\cos\theta'].$$

By virtue of this equality, from (4.14) follows

$$\int_0^\pi\int_0^\pi\frac{A(\rho,\gamma)}{(1-2\rho\cos\gamma+\rho^2)^{\frac{7}{2}}}\sin\theta'd\theta'd\varphi'=0. \tag{4.15}$$

Consider $A(\rho,\gamma)$ on the interval $E=\Big(0\le\theta'\le\frac{\pi}{2};\frac{\pi}{2}\le\varphi'\le\pi\Big)$.

Clearly,

$$-2(1+\rho^2)\sin\theta\sin\theta'\cos\varphi'=2(1+\rho^2)\sin\theta\sin\theta'|\cos\varphi'|>\sin^2\theta\sin^2\theta'\cos^2\varphi'.$$

Next, it can be easily shown that there exists a number $\rho_0>0$ such that on the interval E

$$5\rho-\rho\cos^2\theta\cos^2\theta'-2(1+\rho^2)\cos\theta\cos\theta'>0\quad\text{for}\quad\rho\ge\rho_0.$$

Consequently, on the interval E, for $\rho\ge\rho_0$ we have

$$A(\rho,\gamma)>-2\rho\sin\theta\sin\theta'\cos\theta\cos\theta'\cos\varphi'. \tag{4.16}$$

By (4.16) we obtain

$$I=\int_0^{\frac{\pi}{2}}\int_{\frac{\pi}{2}}^{\pi}\frac{(1-\rho)A(\rho,\gamma)}{(1-2\rho\cos\gamma+\rho^2)^{\frac{7}{2}}}\sin\theta'd\theta'd\varphi'$$
$$>(1-\rho)\int_0^{\frac{\pi}{2}}\int_{\frac{\pi}{2}}^{\pi}\frac{\sin\theta\cos\theta\sin^2\theta'\cos\theta'(-\cos\varphi')}{(1-2\rho\cos\gamma+\rho^2)^{\frac{7}{2}}}d\theta'd\varphi'$$
$$>(1-\rho)^2\cos\theta\int_0^{1-\rho}\int_{\frac{\pi}{2}}^{\pi}\frac{\sin^2\theta'\cos\theta'(-\cos\varphi')}{(1-2\rho\cos\gamma+\rho^2)^{\frac{7}{2}}}d\theta'd\varphi'$$
$$>(1-\rho)^2\cos\theta\cos(1-\rho)\int_0^{1-\rho}\int_{\frac{\pi}{2}}^{\pi}\frac{\sin^2\theta'(-\cos\varphi')}{[(1-\rho)^2+4\rho\sin^2\frac{\gamma}{2}]^{\frac{7}{2}}}d\theta'd\varphi'. \tag{4.17}$$

γ with respect to φ' takes its maximal value when the point $y(1,\theta',\varphi')$ lies on the plane xoz, and $\varphi' = \pi$. Then $\gamma = \theta + \theta'$.

Therefore

$$\begin{aligned}\left[(1-\rho)^2 + 4\rho\sin^2\frac{\gamma}{2}\right]^{\frac{7}{2}} &\leq \left[(1-\rho)^2 + 4\rho\sin^2\frac{\theta+\theta'}{2}\right]^{\frac{7}{2}} \\ &\leq \left[(1-\rho)^2 + 4\rho\sin^2\frac{\theta+(1-\rho)}{2}\right]^{\frac{7}{2}} \\ &\leq \left[(1-\rho)^2 + C\sin^2(1-\rho)\right]^{\frac{7}{2}} < C(1-\rho)^7. \end{aligned} \tag{4.18}$$

The inequalities (4.17) and (4.18) imply

$$\begin{aligned} I &> \frac{C\cos\theta\cos(1-\rho)}{(1-\rho)^5}\int_0^{1-\rho}\theta'^2 d\theta' \int_{\frac{\pi}{2}}^{\pi}(-\cos\varphi')d\varphi' \\ &= \frac{C\sqrt{2\rho-\rho^2}\cos(1-\rho)}{(1-\rho)^2} \to +\infty \quad \text{for} \quad \rho \to 1. \end{aligned} \tag{4.19}$$

Thus, by (4.15) we have

$$\lim_{\rho\to 1}\int_D \frac{(1-\rho)A(\rho,\gamma)}{(1-2\rho\cos\gamma+\rho^2)^{\frac{7}{2}}}\sin\theta' d\theta' d\varphi' = -\infty, \tag{4.20}$$

where

$$D = [0,\pi;0,\pi]\setminus\left[0,\frac{\pi}{2};\frac{\pi}{2},\pi\right].$$

Define now $f(\theta,\varphi)$ as follows:

$$f(\theta,\varphi) = \begin{cases} 0, & \text{in a pole} \\ 0, & \text{when} \quad 0 \leq \theta \leq \pi;\ \pi \leq \varphi \leq 2\pi, \\ 1, & \text{when} \quad 0 \leq \theta \leq \frac{\pi}{2};\ \frac{\pi}{2} \leq \varphi < \pi, \\ -1, & \text{when} \quad (\theta,\varphi) \in D. \end{cases}$$

It is obvious that for this function

$$\overline{\Delta} f(0,0) = 0.$$

However, as follows from (4.19) and (4.20),

$$D_3U(f;\overline{x}) \to +\infty,$$

as $\overline{x}(\rho,\theta,\varphi) \to x(1,0,0)$ along the chosen path.

Theorem 3.4.1 is proved. □

Theorem 3.4.2. *If at a point $x^0 \in S^{k-1}$ there exists a finite $\overline{\Delta}_x f(x^0)$ (see (3.9)), then*

$$D_k U(f;\rho,x) \to \overline{\Delta}_x f(x^0),$$

no matter how a point (ρ,x), $x \in S^{k-1}$ tends to x^0, remaining inside the sphere S^{k-1}.

Proof. Let $\varepsilon > 0$. Choose $\delta > 0$ such (see (4.11)) that

$$|\Delta_\gamma(f;x) - \overline{\Delta}_x f(x^0)| < \varepsilon \tag{4.21}$$

when $0 < \gamma < \delta$, $\rho(x,x^0) < \delta$.

Further, it is easy to check (see the proof of Theorem 3.4.1) that

$$\frac{1}{\rho} D_k U(f;\rho,x) = \frac{2\Gamma\left(\frac{k}{2}\right)}{\rho(k-1)\pi^{\frac{1}{2}}\gamma\left(\frac{k-1}{2}\right)} \int\limits_0^\pi D_k P(\rho,\gamma)\sin^{2\lambda}\gamma \sin^2\frac{\gamma}{2}[\Delta_\gamma(f;x)$$
$$-\overline{\Delta}_x f(x^0)]d\gamma + \overline{\Delta}_x f(x^0) = I_1 + I_2 + \overline{\Delta} f(x^0). \tag{4.22}$$

By Lemma 3.4.2 it can be shown that

$$\lim_{\rho\to 1} I_2 = 0. \tag{4.23}$$

Taking (4.21) into account, we have

$$|I_1| < \varepsilon, \quad \text{when} \quad \rho(x,x^0) < \delta. \tag{4.24}$$

The validity of Theorem 3.4.2 follows from (4.22), (4.23) and (4.24). □

Theorem 3.4.3. *If at a point $x^0 \in S^{k-1}$ there exists a finite $\widetilde{\Delta} f(x^0)$ (see (3.3)), then*

$$\lim_{\rho\to 1} D_k U(f;\rho,x^0) = \widetilde{\Delta} f(x^0).$$

We will prove the theorem for the case $k = 3$, but first we will prove

Lemma 3.4.3. *The following statements are valid:*

1) $\lim\limits_{\rho\to 1} \max\limits_{0<\delta\le\gamma\le\pi} \left|\dfrac{\partial D_3 P(\rho,\gamma)}{\partial\gamma}\right| = 0;$

2) $\int\limits_0^\pi D_3 P(\rho,\gamma)\sin^2\dfrac{\gamma}{2}\sin\gamma d\gamma = 2\rho;$

3) $\int\limits_0^\pi \dfrac{\partial D_3 P(\rho,\gamma)}{\partial\gamma}\sin^4\dfrac{\gamma}{2} d\gamma = -2\rho + o(1)$, *as* $\rho \to 1;$

4) $\int\limits_0^\pi \gamma^4 \left|\frac{\partial D_3 P(\rho,\gamma)}{\partial\gamma}\right| d\gamma = O(1).$

Proof. Statement 1) is obvious. Statement 2) follows from statement 3) and Lemma 3.4.2 ($k = 3$).

For $k = 3$, (4.7) yields

$$D_3P(\rho,\gamma) = -2\rho\frac{P(\rho,\gamma)}{\partial\rho} - \rho^2\frac{\partial^2 P(\rho,\gamma)}{\partial\rho^2} = -4\sum_{n=1}^{\infty} n\Big(n+\frac{1}{2}\Big)P_n(\cos\gamma)\rho^n$$
$$-2\sum_{n=2}^{\infty}(n-1)n\Big(n+\frac{1}{2}\Big)P_n(\cos\gamma)\rho^n. \tag{4.25}$$

By integration by parts, we obtain

$$\int_0^\pi \frac{\partial D_3P(\rho,\gamma)}{\partial\gamma}\sin^4\frac{\gamma}{2}d\gamma = D_3P(\rho,\gamma)\sin^4\frac{\gamma}{2}\Big|_0^\pi - \int_0^\pi D_3P(\rho,\gamma)\sin\gamma\sin^2\frac{\gamma}{2}d\gamma$$
$$= 0(1) - \int_0^\pi D_3P(\rho,\gamma)\sin\gamma\frac{1-\cos\gamma}{2}d\gamma$$
$$= -\frac{1}{2}\int_0^\pi D_3P(\rho,\gamma)[1-P_1(\cos\gamma)]\sin\gamma d\gamma + o(1).$$

Hence by virtue of (4.25) we have

$$\int_0^\pi \frac{\partial D_3P(\rho,\gamma)}{\partial\gamma}\sin^4\frac{\gamma}{2}d\gamma = -3\rho\int_0^\pi [P_1(\cos\gamma)]^2\sin\gamma d\gamma + o(1)$$
$$-3\rho\int_0^\pi \cos^2\gamma\sin\gamma d\gamma = -2\rho + o(1).$$

Let us now prove the validity of statement 4). By Lemma 3.4.1, it can be easily verified that

$$\frac{\partial D_3P(\rho,\gamma)}{\partial\gamma} = \frac{A(\rho,\gamma)}{[(1-\rho)^2 + 4\rho\sin^2\frac{\gamma}{2}]^{\frac{9}{2}}}, \tag{4.26}$$

where all terms in $A(\rho,\gamma)$ with respect to $(1-\rho)$ and $\sin\frac{\gamma}{2}$ are of degree ≥ 4, and $A(\rho,\gamma)$ is divided by $(1-\rho^2)$. Consequently, the numerator $A(\rho,\gamma)$ in the equality (4.26) can be estimated as follows:

$$|A(\rho,\gamma)| \leq (1-\rho)I(\rho,\gamma), \tag{4.27}$$

where $I(\rho,\gamma)$ is a homogeneous polynomial of degree 3 of $(1-\rho,\gamma)$ with a positive coefficient.

Taking into account (4.27) and the inequality $|\sin\gamma| \geq \frac{2}{\pi}|\gamma|$ for $|\gamma| \leq \frac{\pi}{2}$, we obtain

$$\int_0^\pi \gamma^4 \Big|\frac{\partial D_3 P(\rho,\gamma)}{\partial\gamma}\Big| d\gamma \leq (1-\rho)\int_0^\pi \frac{I(\rho,\gamma)\gamma^4 d\gamma}{\Big[(1-\rho)^2 + 4\rho\sin^2\frac{\gamma}{2}\Big]^{\frac{9}{2}}}$$

$$\leq C(1-\rho\int_0^\pi \frac{I(\rho,\gamma)\gamma^4 d\gamma}{[\pi^2(1-\rho)^2 + 4\rho\gamma^2]^{\frac{9}{2}}}. \tag{4.28}$$

Using the substitution $\gamma = (1-\rho)t$ and assuming $\rho > \frac{1}{2}$, from (4.28) we find that

$$\int_0^\pi \Big|\frac{\partial D_3 P(\rho,\gamma)}{\partial\gamma}\Big|\gamma^4 d\gamma < C(1-\rho)\int_0^{\frac{\pi}{1-\rho}} \frac{(1-\rho)^3(1-\rho)^4(1-\rho)\Big(\sum_{v=0}^3 C_v t^v\Big)t^4 dt}{(1-\rho)^9(1+t^2)^{\frac{9}{2}}}$$

$$< C\int_0^\infty \frac{\Big(\sum_{v=0}^3 t^{4+v}\Big)dt}{(1+t^2)^{\frac{9}{2}}} = O(1).$$

Thus statement 4) and Lemma 3.4.3 are proved. □

Proof of Theorem 3.4.3. For the case $k=3$, we have

$$\widetilde{\Delta}_h(f;x^0) = \frac{\frac{1}{4\pi\sin^2\frac{h}{2}}\int_{D^2(x^0;h)} f(y)dS^2(y) - f(x^0)}{\frac{1}{2}\sin^2\frac{h}{2}}$$

$$D_3U(f;\rho,x^0) = -\sum_{n=1}^\infty n(n+1)Y_n(f;x^0)\rho^n = \frac{1}{4\pi}\int_{S^2} D_3P(\rho,\gamma)f(y)dS^2(y).$$

Hence

$$\frac{1}{\rho}D_3U(f;\rho,x^0) = \frac{1}{4\pi\rho}\int_{S^2} D_3P(\rho,\gamma)[f(y)-f(x^0)]dS^2(y)$$

$$= \frac{1}{4\pi\rho}\int_0^\pi D_3P(\rho,\gamma)d\gamma \int_{(x^0,y)=\cos\gamma} [f(y)-f(x^0)]dS^1(y).$$

Using integration by parts and taking into account the statements of Lemma 3.4.3 (assuming $\int_{S^2} f(t)dS^2(t) = 0$), we obtain

$$\frac{1}{\rho}D_3U(f;\rho,x^0) = -\frac{1}{4\pi\rho}\int_0^\pi \frac{\partial D_3P(\rho,\gamma)}{\partial\gamma}d\gamma \int_{D^2(x^0;\gamma)} [f(y)-f(x^0)]dS^2(y)$$

$$= -\frac{1}{2\rho}\int_0^\pi \frac{\partial D_3 P(\rho,\gamma)}{\partial\gamma}\sin^4\frac{\gamma}{2}\left\{\frac{\frac{1}{4\pi\sin^2\frac{\gamma}{2}}\int\limits_{D^2(x^0;\gamma)}[f(y)-f(x^0)]dS^2(y)}{\frac{1}{2}\sin^2\frac{\gamma}{2}}\right\}d\gamma$$

$$= -\frac{1}{2\rho}\int_0^\pi \frac{\partial D_3 P(\rho,\gamma)}{\partial\gamma}\sin^4\frac{\gamma}{2}$$

$$\times\left\{\frac{\frac{1}{4\pi\sin^2\frac{\gamma}{2}}\int\limits_{D^2(x^0;\gamma)}[f(y)-f(x^0)]dS^2(y)}{\frac{1}{2}\sin^2\frac{\gamma}{2}} - \widetilde{\Delta} f(x^0)\right\}d\gamma + \widetilde{\Delta} f(x^0).$$

Using Lemma 3.4.3 as before, from the latter equality we easily establish the validity of Theorem 3.4.3. □

Our next statement is also easy to prove.

Theorem 3.4.4. *If at a point $x^0 \in S^{k-1}$ there exists a finite $\widetilde{\Delta}_x f(x^0)$, then*

$$D_k U(f;\rho,x) \to \widetilde{\Delta}_x f(x^0),$$

no matter how the point (ρ,x) tends to x^0, remaining inside the sphere S^{k-1}.

Remark. Theorems 3.4.1 and 3.4.2 are the corollaries of Theorems 3.4.3 and 3.4.4, respectively.

3.5 The Boundary Properties of the Integral $D_k^r U(f;\rho,\theta_1,\theta_2,\ldots,\theta_{k-2},\varphi)$, $r \in N$

In this section we prove the theorems on the boundary properties of the integral $D_k^r U(f;\rho,\theta_1,\theta_2,\ldots,\theta_{k-2},\varphi)$, $r \in N$ when on the unit sphere S^{k-1} the density of the Poisson integral for a ball has a generalized Laplace operator of any order $r \in N$.

Lemma 3.5.1. *If $P(\rho,\gamma)$ is the Poisson kernel, then*

$$D_k^r P(\rho,\gamma) = \frac{A_r(\rho,\gamma)}{[(1-\rho)^2 + 4\rho\sin^2\frac{\gamma}{2}]^{\lambda+2r+1}}, \tag{5.1}$$

where all terms in $A_r(\rho,\gamma)$ with respect to $(1-\rho)$ and $\sin\frac{\gamma}{2}$ are of degree $\geq 2r+1$, and $A_r(\rho,\gamma)$ is divided by $(1-\rho^2)$.

Proof. Let ([3])

$$P(\rho,\gamma) = (1-\rho^2)[\Delta(\rho,\gamma)]^{-(\lambda+1)}, \tag{5.2}$$

where

$$\Delta(\rho,\gamma) = (1-\rho)^2 + 4\rho\sin^2\frac{\gamma}{2}.$$

By (4.1) and (5.2), we obtain

$$D_k P(\rho,\gamma) = -\frac{1}{\rho^{k-3}}\cdot\frac{\partial}{\partial\rho}\Big[\rho^{k-1}\frac{\partial P(\rho,\gamma)}{\partial\rho}\Big]$$
$$= -\frac{1}{\rho^{k-3}}\cdot\frac{\partial}{\partial\rho}\rho^{k-1}\Big\{-2\rho[\Delta(\rho,\gamma)]^{-(\lambda+1)}$$
$$-(\lambda+1)(1-\rho^2)[\Delta(\rho,\gamma)]^{-(\lambda+2)}\frac{\partial\Delta(\rho,\gamma)}{\partial\rho}\Big\}$$
$$= \frac{1}{\rho^{k-3}}\cdot\frac{\partial}{\partial\rho}\{B_1(\rho,\gamma)[\Delta(\rho,\gamma)]^{-(\lambda+2)}\}, \tag{5.3}$$

where

$$B_1(\rho,\gamma) = \rho^{k-1}\Big[2\rho\Delta(\rho,\gamma) + (\lambda+1)(1-\rho^2)\frac{\partial\Delta(\rho,\gamma)}{\partial\rho}\Big].$$

Clearly, all terms in $B_1(\rho,\gamma)$ with respect to $(1-\rho)$ and $\sin\frac{\gamma}{2}$ are of degree ≥ 2. From (5.3) it follows that

$$D_k P(\rho,\gamma) = \frac{1}{\rho^{k-3}}\Big\{\frac{\partial B_1(\rho,\gamma)}{\partial\rho}[\Delta(\rho,\gamma)]^{-(\lambda+2)}$$
$$-(\lambda+2)[\Delta(\rho,\gamma)]^{-(\lambda+3)}\frac{\partial\Delta(\rho,\gamma)}{\partial\rho}\cdot B_1(\rho,\gamma)\Big\}$$
$$= A_1(\rho,\gamma)[\Delta(\rho,\gamma)]^{-(\lambda+3)} = (1-\rho^2)D_k[\Delta(\rho,\gamma)]^{-(\lambda+1)}], \tag{5.4}$$

where

$$A_1(\rho,\gamma) = \frac{1}{\rho^{k-3}}\Big[\frac{\partial B_1(\rho,\gamma)}{\partial\rho}\cdot\Delta(\rho,\gamma) - (\lambda+2)B_1(\rho,\gamma)\frac{\partial\Delta(\rho,\gamma)}{\partial\rho}\Big].$$

It is not difficult to see that all terms in $A_1(\rho,\gamma)$ with respect to $(1-\rho)$ and $\sin\frac{\gamma}{2}$ are of degree ≥ 3, and as it follows from (5.4), $A_1(\rho,\gamma)$ is divided by $(1-\rho^2)$.

Assume now that this is the case for some $r\in N$. Then (5.1) yields

$$D_k^{r+1}P(\rho,\gamma) = D_k\{A_r(\rho,\gamma)[\Delta(\rho,\gamma)]^{-(\lambda+2r+1)}\}$$
$$= -\frac{1}{\rho^{k-3}}\cdot\frac{\partial}{\partial\rho}\Big\{\rho^{k-1}\frac{\partial}{\partial\rho}A_r(\rho,\gamma)[\Delta(\rho,\gamma)]^{-(\lambda+2r+1)}\Big\}$$
$$= -\frac{1}{\rho^{k-3}}\cdot\frac{\partial}{\partial\rho}\rho^{k-1}\Big\{\frac{\partial A_r(\rho,\gamma)}{\partial\rho}[\Delta(\rho,\gamma)]^{-(\lambda+2r+1)}$$
$$-(\lambda+2r+1)[\Delta(\rho,\gamma)]^{-(\lambda+2r+1)}\frac{\partial\Delta(\rho,\gamma)}{\partial\rho}\cdot A_r(\rho,\gamma)\Big\}$$
$$= -\frac{1}{\rho^{k-3}}\cdot\frac{\partial}{\partial\rho}\Big\{B_r(\rho,\gamma)[\Delta(\rho,\gamma)]^{-(\lambda+2r+2)}\Big\},$$

where

$$B_r(\rho,\gamma) = \rho^{k-1}\Big[\frac{\partial A_r(\rho,\gamma)}{\partial\rho}\cdot\Delta(\rho,\gamma) - (\lambda+2r+1)A_r(\rho,\gamma)\frac{\partial\Delta(\rho,\gamma)}{\partial\rho}\Big].$$

As is easily seen, all terms in $B_r(\rho,\gamma)$ with respect to $(1-\rho)$ and $\sin\frac{\gamma}{2}$ are of degree $\geq 2r+2$.

Furthermore,

$$D_k^{r+1}P(\rho,\gamma) = -\frac{1}{\rho^{k-3}}\frac{\partial}{\partial\rho}\{B_r(\rho,\gamma)[\Delta(\rho,\gamma)]^{-(\lambda+2r+2)}\}$$
$$= -\frac{1}{\rho^{k-3}}\Big\{\frac{\partial B_r(\rho,\gamma)}{\partial\rho}[\Delta(\rho,\gamma)]^{-(\lambda+2r+2)}$$
$$-(\lambda+2r+2)[\Delta(\rho,\gamma)]^{-(\lambda+2r+3)}\frac{\partial\Delta(\rho,\gamma)}{\partial\rho}\cdot B_r(\rho,\gamma)\Big\}$$
$$= A_{r+1}(\rho,\gamma)[\Delta(\rho,\gamma)]^{-(\lambda+2r+3)} = (1-\rho^2)D_k^{r+1}[\Delta(\rho,\gamma)]^{-(\lambda+1)}], \tag{5.5}$$

where

$$A_{r+1}(\rho,\gamma) = -\frac{1}{\rho^{k-3}}\Big\{\frac{\partial B_r(\rho,\gamma)}{\partial\rho}\cdot\Delta(\rho,\gamma) - (\lambda+2r+2)B_r(\rho,\gamma)\frac{\partial\Delta(\rho,\gamma)}{\partial\rho}\Big\}.$$

Hence it is clear that all terms in $A_{r+1}(\rho,\gamma)$ with respect to $(1-\rho)$ and $\frac{\gamma}{2}$ are of degree $\geq 2r+3$ and, as follows from (5.5), $A_{r+1}(\rho,\gamma)$ is divided by $(1-\rho^2)$.

Thus Lemma 3.5.1 is proved. □

The property $A_r(\rho,\gamma)$ implies

$$|A_r(\rho,\gamma)| \leq (1-\rho)I_r(\rho,\gamma), \tag{5.6}$$

where $I_r(\rho,\gamma)$ is a homogeneous polynomial of degree $2r$ of $(1-\rho,\gamma)$ with a positive coefficient.

Lemma 3.5.2. *For any $r\in N$, the following statements are valid:*

1) $\int\limits_0^{\pi} |D_k^r P(\rho,\gamma)|\gamma^{2r+k-2}d\gamma = O(1)$,

2) $\lim\limits_{\rho\to 1}\max\limits_{0<\delta\leq\gamma\pi} |D_k^r P(\rho,\gamma)| = 0$.

Proof. Taking into account (5.6) and the inequality $|\sin\gamma| \geq \frac{2}{\pi}|\gamma|$ for $|\gamma|\leq\frac{\pi}{2}$, we have

$$\int\limits_0^{\pi} |D_k^r P(\rho,\gamma)|\gamma^{2r+2\lambda}d\gamma \leq (1-\rho)\int\limits_0^{\pi}\frac{I_r(\rho,\gamma)\gamma^{2r+2\lambda}d\gamma}{\Big[(1-\rho)^2+4\rho\sin^2\frac{\gamma}{2}\Big]^{2r+\gamma+1}}$$
$$\leq C(1-\rho)\int\limits_0^{\pi}\frac{I_r(\rho,\gamma)\gamma^{2r+2\lambda}}{\Big[\pi^2(1-\rho)^2+4\rho\gamma^2\Big]^{2r+\gamma+1}}. \tag{5.7}$$

Using the substitution $\gamma=(1-\rho)t$ and assuming $\rho>\frac{1}{2}$, from (5.7) we obtain

$$\int\limits_0^{\pi} |D_k^r P(\rho,\gamma)|\gamma^{2r+2\lambda}d\gamma$$

$$< C(1-\rho)\int\limits_0^{\frac{\pi}{1-\rho}} \frac{(1-\rho)^{2r}(1-\rho)^{2r+2\lambda}(1-\rho)\Big(\sum_{v=0}^{2r} C_v t^v\Big) t^{2r+2\lambda}dt}{(1-\rho)^{4r+2\lambda+2}(1+t^2)^{2r+\lambda+1}}$$

$$< C\int\limits_0^{\infty} \frac{\Big(\sum_{v=0}^{2r} t^{2r+2\lambda+v}\Big)dt}{(1+t^2)^{2r+\lambda+1}} = O(1).$$

Thus statement 1) is proved. Statement 2) is obvious.

Lemma 3.5.2 is proved. □

Lemma 3.5.3 ([69], p. 287). *If from the conditions $g(x)=0$ and $\overline{\Delta}^v g(x)=0$, $v=1,2,\dots,r$, follows the equality*

$$\lim_{\rho\to 1-} D_k^r U(g;\rho,x)=0,$$

then the equality $\overline{\Delta}^r g(x)=S$ implies that

$$\lim_{\rho\to 1-} D_k^r U(g;\rho,x)=S.$$

Proof. Let us first prove that there exists a finite sum of spherical harmonics $T(y)=\sum\limits_{j=0}^{r} a_j P_j^\lambda([x,y])$ such that

$$\Delta^v T(x)=\overline{\Delta}^v f(x), \quad v=0,1,2,\dots,r.$$

Indeed, since the operators Δ^v and $\overline{\Delta}^v$ are linear,

$$\overline{\Delta}^v T(y)=\sum_{j=0}^{r} a_j \Delta^v P_j^\lambda([x,y])=\Delta^v T(y),$$

we can choose a_j such that

$$\left.\begin{array}{l} \sum\limits_{j=0}^{r} a_j P_j^\lambda(1)=\overline{\Delta}^0 f(x), \\ \sum\limits_{j=1}^{n} [-j(j+k-2)]^v a_j P_j^\lambda(1)=\overline{\Delta}^v f(x), \quad v=1,2,\dots,r \end{array}\right\}.$$

This system has a unique solution because its (Vandermonde) determinant is not equal to zero.

Let now $g(t)=f(t)-T(t)$, where $T(x)=f(x)$ and $\Delta^\nu T(x)=\overline{\Delta}^v f(x)$, $v=1,2,\dots,r$ ($\Delta^r T(x)=S$).

Moreover,

$$U(g;\rho,x)=U(f;\rho,x)-U(T;\rho,x)=U(f;\rho,x)-\sum_{v=0}^{r} a_v P_v^\lambda(\cos\gamma)\rho^v,$$

whence

$$D_k^r U(g;\rho,x) = D_k^r U(f;\rho,x) - \sum_{v=0}^{r} a_\nu D_k^r P_k^\lambda(\cos\gamma)\rho^v.$$

This implies that

$$\lim_{\rho\to 1-} D_k^r U(g;\rho,x) = \lim_{\rho\to 1} D_k^r U(f;\rho,x) - D_k^r T(x) = \lim_{\rho\to 1} D_k^r U(f;\rho,x) - S = 0.$$

Lemma 3.5.3 is proved. □

Theorem 3.5.1. *If at the point $x^0(1,\theta_1^0,\theta_2^0,\ldots,\theta_{k-2}^0,\varphi^0)$ there exists a finite $\overline{\Delta}^r f(x^0)$, then*

$$\lim_{\rho\to 1-} D_k^r U(f;\rho,x^0) = \overline{\Delta}^r f(x^0).$$

Proof. Assuming that x^0 is the initial point, we have

$$\begin{aligned} D_k^r U(f;\rho,x^0) &= \frac{\Gamma\left(\frac{k}{2}\right)}{2\pi^{\frac{k}{2}}} \int\limits_{S^{k-1}} D_k^r P(\rho,\gamma) f(t)\,dS^{k-1}(t) \\ &= \frac{\Gamma\left(\frac{k}{2}\right)}{2\pi^{\frac{k}{2}}} \int\limits_0^\pi D_k^r P(\rho,\gamma)\,d\gamma \int\limits_{C^{k-1}(x^0;\gamma)} f(t)\,dS^{k-2}(t) \\ &= \frac{\Gamma\left(\frac{k}{2}\right)}{\pi^{\frac{1}{2}}\Gamma\left(\frac{k-1}{2}\right)} \int\limits_0^\pi D_k^r P(\rho,\gamma)\sin^{k-2}\gamma\,d\gamma \\ &\quad\times \frac{1}{|S^{k-2}|\sin^{k-2}\gamma} \int\limits_{C^{k-1}(x^0;\gamma)} f(t)\,dS^{k-2}(t). \end{aligned} \tag{5.8}$$

By Lemma 3.5.3, it can be assumed that $f(x^0) = \overline{\Delta}^1 f(x^0) = \cdots = \overline{\Delta}^r f(x^0) = 0$. Let $\varepsilon > 0$ and choose $\delta > 0$ such that

$$\frac{1}{|S^{k-2}|\sin^{k-2}\gamma}\left|\int\limits_{C^{k-2}(x^0;\gamma)} f(t)\,dS^{k-2}(t)\right| < \varepsilon\gamma^{2r} \quad\text{for}\quad 0<\gamma<\delta. \tag{5.9}$$

Then from (5.8) we have

$$\begin{aligned} D_k^r U(f;\rho,x^0) &= \frac{\Gamma\left(\frac{k}{2}\right)}{\sqrt{\pi}\,\Gamma\left(\frac{k-1}{2}\right)}\Bigg\{\int\limits_0^\delta D_k^r P(\rho,\gamma)\sin^{k-2}\gamma\,d\gamma \\ &\quad\times \frac{1}{|S^{k-2}|\sin^{k-2}\gamma} \int\limits_{C^{k-2}(x^0;\gamma)} f(t)\,dS^{k-2}(t) \end{aligned}$$

$$+\int_{\delta}^{\pi} D_k^r P(\rho,\gamma)\sin^{k-2}\gamma d\gamma \frac{1}{|S^{k-2}|\sin^{k-2}\gamma}\int_{C^{k-2}(x^0;\gamma)} f(t)dS^{k-2}(t)\Bigg\} = I_1+I_2.$$

By virtue of (5.9) and statement 1) of Lemma 3.5.2, we obtain

$$|I_1| < C\varepsilon \int_0^{\pi} |D_k^r P(\rho,\gamma)|\gamma^{2r+k-2}d\gamma < C\varepsilon. \tag{5.10}$$

Furthermore, taking into account statement 2) of Lemma 3.5.2, we obtain

$$|I_2| \le C\max|D_k^r P(\rho,\gamma)| \int_{S^{k-1}} |f(t)|dS^{k-1}(t). \tag{5.11}$$

By (5.10) and (5.11) we conclude that Theorem 3.5.1 is valid. □

Lemma 3.5.4. *Given finite functions $\alpha_i(x)$, $i=0,1,2,3\dots,r$, there exists a function*

$$T(x,t)=\sum_{v=0}^{r} a_v(x)P_v^{\lambda}([x,t])$$

possessing the properties

$$T(x,x)=\alpha_0(x), \quad D_k^v T(x,x)=\alpha_v(x), \quad v=1,2,\dots,r.$$

This lemma can be proved analogously to Lemma 3.5.3.

Theorem 3.5.2. *If at a point $x^0\in S^{k-1}$ there exists a finite $\overline{\Delta}_x^r(x^0)$, then*

$$\lim_{(\rho,x)\to(1,x^0)} D_k^r U(f;\rho,x)=\overline{\Delta}_x^r f(x^0).$$

Proof. We have

$$D_k^r U(f;\rho,x)=\frac{\Gamma\left(\frac{k}{2}\right)}{2\pi^{\frac{k}{2}}}\int_{S^{k-1}} D_k^r P(\rho,\gamma)f(t)dS^{k-1}(t)$$

$$=\frac{\Gamma\left(\frac{k}{2}\right)}{2\pi^{\frac{k}{2}}}\int_0^{\pi} D_k^r P(\rho,\gamma)d\gamma \int_{C^{k-2}(x;\gamma)} f(t)dS^{k-2}(t)$$

$$=\frac{\Gamma\left(\frac{k}{2}\right)}{\Gamma\left(\frac{k-1}{2}\right)\sqrt{\pi}}\int_0^{\pi} D_k^r P(\rho,\gamma)\sin^{k-2}\gamma \frac{1}{|S^{k-2}|\sin^{k-2}\gamma}\int_{C^{k-2}(x;\gamma)} f(t)dS^{k-2}(t).$$

According to Lemmas 3.5.3 and 3.5.4, for any $\varepsilon > 0$ we can choose $\delta > 0$ such that

$$\frac{1}{|S^{k-2}|\sin^{k-2}\gamma}\left|\int\limits_{C^{k-2}(x;\gamma)} f(t)dS^{k-2}(t)\right| < \varepsilon\gamma^{2r} \quad \text{for} \quad 0 < \gamma < \delta \quad \text{and} \quad \rho(x,x^0) < \delta.$$

Our further reasoning will be the same as in proving Theorem 3.5.1. □

Lemma 3.5.5. *For any $r \in N$, the following statements are valid:*
1) $\int\limits_0^\pi \left|\frac{\partial D_k^r P(\rho,\gamma)}{\partial\gamma}\right|\gamma^{2r+k-1}d\gamma = O(1)$,
2) $\lim\limits_{\rho\to 1}\max\limits_{0<\delta\le r\le\pi}\left|\frac{\partial D_k^r P(\rho,\gamma)}{\partial\gamma}\right| = 0.$

Proof. From (5.1) it follows that

$$\frac{\partial D_k^r P(\rho,\gamma)}{\partial\gamma} = \frac{T_r(\rho,\gamma)}{\left[(1-\rho^2)+4\rho\sin^2\frac{\gamma}{2}\right]^{\gamma+2r+2}},$$

where all terms in $T_r(\rho,\gamma)$ with respect to $(1-\rho)$ and $\sin\frac{\gamma}{2}$ are of degree $\ge 2r+2$, and $T_r(\rho,\gamma)$ is divided by $(1-\rho^2)$. Therefore

$$|T_r(\rho,\gamma)| \le (1-\rho)I_{r+1}(\rho,\gamma), \tag{5.12}$$

where $I_{r+1}(\rho,\gamma)$ is a homogeneous polynomial of degree $2r+1$ of $(1-\rho,\gamma)$, and all its coefficients are positive.

Using the substitution $\gamma = (1-\rho)t$ and taking (5.12) into account, for $\rho > \frac{1}{2}$ we have

$$\int\limits_0^\pi \left|\frac{\partial D_k^r P(\rho,\gamma)}{\partial\gamma}\right|\gamma^{2r+k-1}d\gamma \le C(1-\rho)\int\limits_0^\pi \frac{I_{r+1}(\rho,\gamma)\gamma^{2r+k-1}d\gamma}{\left[(1-\rho^2)+4\rho\sin^2\frac{\gamma}{2}\right]^{\gamma+2r+2}}$$

$$\le C(1-\rho)\int\limits_0^{\frac{\pi}{1-\rho}} \frac{(1-\rho)^{2r+1}(1-\rho)^{2r+k-1}(1-\rho)\left(\sum\limits_{v=0}^{2r+1}C_v t^v\right)t^{2r+k-1}}{(1-\rho)^{4r+2\lambda+4}(1+t^2)^{\lambda+2r+2}}dt$$

$$= C\int\limits_0^{\frac{\pi}{1-\rho}} \frac{\left(\sum\limits_{v=0}^{2r+1}t^{2r+2\lambda+1+v}\right)dt}{(1+t^2)^{2r+\lambda+2}} = O(1).$$

Statement 2) is obvious.
Lemma 3.5.5 is proved. □

Theorem 3.5.3. *If at a point $x^0 \in S^{k-1}$ there exists a finite $\widetilde{\Delta}^r f(x^0)$, then*

$$\lim_{\rho\to 1-} D_k^r U(f;\rho,x^0) = \widetilde{\Delta}^r f(x^0).$$

Proof. Using integration by parts and assuming that $\int\limits_{S^{k-1}} f(t)dS^{k-1}(t) = 0$, we obtain

$$D_k^r U(f;\rho,x^0) = \frac{\Gamma\left(\frac{k}{2}\right)}{2\pi^{\frac{k}{2}}} \int\limits_0^\pi D_k^r P(\rho,\gamma)d\gamma \int\limits_{C^{k-2}(x^0;\gamma)} f(t)dS^{k-2}(t)$$

$$= -\frac{\Gamma\left(\frac{k}{2}\right)}{2\pi^{\frac{k}{2}}} \int\limits_0^\pi \frac{\partial D_k^r P(\rho,\gamma)}{\partial\gamma} d\gamma \int\limits_{D^{k-1}(x^0;\gamma)} f(t)dS^{k-1}(t)$$

$$= -\frac{\Gamma\left(\frac{k}{2}\right)}{2\pi^{\frac{k}{2}}} \int\limits_0^\pi \frac{\partial D_k^r P(\rho,\gamma)}{\partial\gamma} |D^{k-1}(x^0;\gamma)| d\gamma \frac{1}{|D^{k-1}(x^0;\gamma)|} \int\limits_{D^{k-1}(x^0;\gamma)} f(t)dS^{k-1}(t).$$

Further, using Lemmas 3.5.3 and 3.5.5 and arguing as in proving Theorem 3.5.1, we obtain the proof of Theorem 3.5.3. □

In the same manner we prove

Theorem 3.5.4. *If at a point $x^0 \in S^{k-1}$ there exists a finite $\widetilde{\Delta}_x^r f(x^0)$, then*

$$\lim_{(\rho,x)\to(1,x^0)} D_k^r U(f;\rho,x) = \widetilde{\Delta}_x^r f(x^0).$$

Remark. Theorems 3.5.1 and 3.5.2 are the corollaries of Theorems 3.5.3 and 3.5.4, respectively.

The boundary properties of first and second order partial derivatives of the Poisson integral for $U(f;\rho,\theta,\varphi)$ e are thoroughly investigated by O.P. Dzagnidze in [18–25] for the case $k = 3$.

3.6 The Dirichlet Problem for a Ball

The Dirichlet problem for a ball is formulated as follows: Given a function $f(x)$ on S^{k-1}, find in V^k a harmonic function function $U(\rho,x)$, $0 < \rho < 1$, $x \in S^{k-1}$, which on S^{k-1} the value $f(x)$.

Theorems 3.2.1 and 3.2.2 in terms of the Dirichlet problem can be formulated as follows.

Theorem 3.6.1. *If $f(x)$ is continuous on S^{k-1}, then the Poisson integral $U(f;\rho,x)$ (which is harmonic in the ball V^k) is a solution of the Dirichlet problem in a sense that for all $\overline{x} = (1,\overline{\theta}_1,\overline{\theta}_2,\dots,\overline{\theta}_{k-2},\varphi) \in S^{k-1}$,*

$$U(f;\rho,x) \to f(\overline{x}),$$

no matter how the point $(\rho,x) = (\rho,\theta_1,\theta_2,\dots,\theta_{k-2},\varphi) \in V^k$ tends to $\overline{x}$, remaining inside the sphere S^{k-1}.

Theorem 3.6.2. *If $f(x) \in L(S^{k-1})$, then the Poisson integral $U(f;\rho,x)$, being a harmonic function in V^k, has almost everywhere on S^{k-1} boundary values along all non-tangential paths to the sphere S^{k-1}, which coincide with the values of f.*

In this section the Dirichlet problem is solved for the case where the boundary function $f(x)$ is measurable and finite almost everywhere on S^{k-1} (N.N. Luzin's formulation). To solve the problem, we need an analogue of Luzin's theorem on the existence of a primitive function for a function of many variables. The proof will be given for a function of two variables.

In [14], A.G. Jvarsheishvili generalized Luzin's theorem to a function of two variables. He proved that if $P(x,y)$ and $Q(x,y)$ are arbitrary measurable and almost everywhere finite functions on $R_0 = [0,1;0,1]$, then there exists a continuous function $F(x,y)$ such that almost everywhere on R_0, $dF(x,y) = P(x,y)dx + Q(x,y)dy$. This problem was posed by G.P. Tolstov [76].

In this section we obtain an analogue of this theorem for a derivative of the function of two variables $f(x,y)$ at the point $\mathbf{P}(x,y)$ and define it as follows:

$$\overline{\Delta} f(x,y) = \lim_{r\to 0} \Delta_r(f;\mathbf{P}) = \lim_{r\to 0} \frac{\frac{1}{2\pi r}\int_{C(P;r)} f(t,\tau)dS(t,\tau) - f(x,y)}{\frac{1}{4}r^2},$$

where $C(\mathbf{P};r)$ is a circumference of radius r, with center at the point $\mathbf{P}(x,y)$.

Theorem 3.6.3 ([84]). *Let $F(x,y)$ be an arbitrary measurable and almost everywhere finite function on R_0. Then there exists a continuous function $F(x,y)$ such that*

$$\overline{\Delta} F(x,y) = f(x,y)$$

almost everywhere on R_0.

First we need to prove a few lemmas.

Lemma 3.6.1. *For any $M > 0$ and for a real number a, there exists, on R_0, a continuous function $F(x,y)$, such that*
1) $F(x,y) = 0$, *when* $(x,y) \in \partial R_0$,
2) $|F(x,y)| \le M^2$, $(x,y) \in R_0$,
3) $\overline{\Delta} F(x,y) = a$ *almost for all* $(x,y) \in R_0$.

Proof. Let $\Phi(x,y) = \frac{a}{4}(x^2+y^2)$, $(x,y) \in R_0$.
It is obvious that

$$\overline{\Delta}\Phi(x,y) = a.$$

By A.G. Jvarsheishvili's Lemma 2 ([14], p. 16), we can construct a stepwise function $S(x,y)$ such that $S(x,y) = \Phi(x,y)$, when $(x,y) \in \partial R_0$, and for all $(x,y) \in R_0$ we have

$$|\Phi(x,y) - S(x,y)| \le M^2.$$

Denote
$$F(x,y)=\Phi(x,y)-S(x,y).$$

It is clear that $F(x,y)$ satisfies all the required conditions.
The lemma is proved. □

Lemma 3.6.2. *Let $G=\cup_{k=1}^{n} r_k\subset R_0$; $r_k=(\alpha_k,\beta_k;\gamma_k,\delta_k)$; $r_i\cap r_j=\varnothing$, $i\neq j$. For any numbers a_k, $k=1,2,\dots,n$ and $M>0$, there exists a continuous function $F(x,y)$ such that:*

1) $F(x,y)=0$; $(x,y)\in R_0-G$;
2) $|F(x,y)|\le M^2$, $(x,y)\in R_0$;
3) $\overline{\Delta}F(x,y)=a_k$ *almost everywhere on* r_k, $k=1,2,\dots,n$;
4) $\overline{\Delta}F(x,y)=0$ *almost everywhere on* R_0-G.

Proof. By Lemma 3.6.1, for each r_k we construct a continuous function $F_k(x,y)$ such that
$$F_k(x,y)=0,\quad (x,y)\in R_0-r_k;\quad |F_k(x,y)|\le M^2,\quad (x,y)\in R_0$$
and almost everywhere on r_k,
$$\overline{\Delta}F_k(x)=a_k.$$

It is obvious that the function
$$F(x,y)=\sum_{k=1}^{n}F_k(x,y)$$
satisfies all the conditions of the lemma. □

Lemma 3.6.3. *Let $f(x,y)$ be the continuous on R_0 function. Then for any $\varepsilon>0$, there exists a continuous function $F(x,y)$ such that*

1) $F(x,y)=0$, *when* $(x,y)\in\partial R_0$,
2) $|F(x,y)|\le\varepsilon^2$ *for all* $(x,y)\in R_0$;
3) *almost everywhere on* R_0
$$|\overline{\Delta}F(x,y)-f(x,y)|<\varepsilon^2.$$

Proof. For $\varepsilon>0$, we choose $\delta(\varepsilon)>0$ such that
$$\max_{\rho(\mathbf{P}';\mathbf{P}'')<\delta}|f(\mathbf{P}')-f(\mathbf{P}'')|<\varepsilon^2.$$

By the straight lines, parallel to the coordinate axes, we divide R_0 into intervals $r_1,r_2,\dots,r_n$ so that $d(r_k)<\delta$, $k=1,2,\dots,n$. We denote by $\mathbf{P}_k$ the center of an interval r_k and assume $a_k=f(\mathbf{P}_k)$.

By Lemma 3.6.2, we can construct $F_k(x,y)$ such that $F_k(x,y)=0$ for all $(x,y)\in R_0-r_k$, $|F_k(x,y)|<\varepsilon^2$ for $(x,y)\in R_0$, and $\overline{\Delta}F_k(x,y)=a_k$ almost everywhere on r_k.

It is obvious that if

$$F(x,y)=\sum_{k=1}^{n}F_k(x,y),$$

then $|F(x,y)|<\varepsilon^2$, $(x,y)\in R_0$ and $\overline{\Delta}F(x,y)=a_k$ almost everywhere on r_k.

Therefore

$$|\overline{\Delta}F(x,y)-f(x,y)|<\varepsilon^2$$

almost everywhere on R_0.

The lemma is proved. □

Proof of Theorem 3.6.3. By Luzin's property (C), there exists a sequence of perfect sets $\{P_n\}$ such that $P_i\cap P_j=\varnothing$, $i\neq j$,

$$R_0=H\cup\left(\bigcup_{n=1}^{\infty}P_n\right),\quad mH=0,$$

and the function $f(x,y)$ is continuous on each set P_n. According to the Brower-Urison theorem ([118], p.133), there exists a continuous function $f_n(x,y)$ on R_0 such that $f_n(x,y)=f(x,y)$ when $(x,y)\in P_n$. Let us cover each P_k by the systems $H_k^{(m)}$ consisting of a finite number of non-overlapping segments so that for any k and m each point P_k would lie strictly inside one of the segments forming $H_k^{(m)}$, and so that

$$H_k^{(m+1)}\subset H_k^{(m)}\quad(m=1,2,3,\dots),$$

$$P_k=\bigcap_{m=1}^{\infty}H_k^{(m)},$$

$$H_{k_1}^{(m)}\cap H_{k_2}^{(m)}=\varnothing.\ \ k_1\neq k_2,\quad 1\le k_1,\quad k_2\le m.$$

Assume

$$g_m(x,y)=\begin{cases}f_k(x,y), & \text{when}\ \ (x,y)\in H_k^{(m)},\quad 1\le k\le m,\\ 0, & \text{when}\ \ (x,y)\overline{\in}\bigcap\limits_{m=1}^{m}H_k^{(m)}.\end{cases}$$

$g_m(x,y)$ is a piecewise continuous function on R_0. It is not difficult to show that if $(x_0,y_0)\in\theta=\bigcap\limits_{k=1}^{\infty}P_k$, then

$$\lim_{m\to\infty}g_m(x_0,y_0)=f(x_0,y_0).\tag{6.1}$$

Putting

$$\Psi_1(x,y)=g_1(x,y),$$

$$\Psi_m(x,y) = g_m(x,y) - g_{m-1}(x,y), \quad m > 1,$$

by (6.1) we have

$$f(x,y) = \sum_{m=1}^{\infty} \Psi_m(x,y).$$

Let us consider the set

$$\mathbf{E}_1 = H_1^{(1)}, \quad \mathbf{E}_m = H_m^{(m)} \cup \left\{ \bigcup_{k=1}^{m-1} (H_k^{(m-1)} - H_k^{(m)}) \right\}.$$

For $(x,y) \in C\mathbf{E}_m$, we have

$$\Psi_m(x,y) = 0. \tag{6.2}$$

Let

$$\Theta_m = \bigcup_{k=1}^{m} P_k.$$

Then

$$\Theta_{m-1} \cap \overline{\mathbf{E}}_m = \varnothing.$$

Hence

$$\rho_m = \rho(\Theta_{m-1}, \mathbf{E}_m) > 0; \quad \lim_{m\to\infty} \rho_m = 0. \tag{6.3}$$

Enumerate arbitrarily the rectangles contained in $\mathbf{E}_1$ and assume $\mathbf{E}_1 = \{\Delta_1, \Delta_2, \dots, \Delta_{v_1}\}$. Then divide $\overline{\mathbf{E}}_2$ into a finite number of rectangles Δ_s, $v_1 < s \leq v_2$ and so on. Generally speaking, $\overline{\mathbf{E}}_m$ is divided into a finite number of rectangles Δ_s, $v_{m-1} < s \leq v_m$.

By virtue of (6.3), if $(x,y) \in \Theta_{m-1}$, $(t,\tau) \in \Delta_s$, $v_{m-1} < s \leq v_m$, then

$$\rho[(x,y),(t,\tau)] = \sqrt{(x-t)^2 + (y-\tau)^2} \geq \rho_m.$$

Assume

$$\varphi_s(x,y) = \begin{cases} \Psi_m(x,y), & \text{when} \quad (x,y) \in \Delta_s, \quad v_{m-1} < s \leq v_m, \\ 0, & \text{when} \quad (x,y) \overline{\in} \Delta_s, \quad v_{m-1} < s \leq v_m. \end{cases}$$

It is easy to show that on Θ,

$$f(x,y) = \sum_{s=1}^{\infty} \varphi_s(x,y).$$

Note that $\varphi_s(x,y)$ are piecewise continuous functions on R_0. By the latter fact and Lemma 3.6.3, for $\frac{r_1}{2}$, $r_1 > 0$ there exists a continuous function $F_1(x,y)$ on R_0 such that for all $(x,y) \in R_0$ we have

$$|F_1(x,y)| < \frac{r_1^2}{2};$$

$F_1(x,y)=0$ when $(x,y)\in R_0-\Delta_1$ and almost everywhere on R_0,

$$|\overline{\Delta}F_1(x,y)-\varphi_1(x,y)|<\frac{r_1^2}{2\cdot 2}.$$

By Egorov's theorem and Tolstov's remark ([57], p.310), for the number $\frac{1}{2}$, there exists a closed set W_1, $|R_0-W_1|<\frac{1}{2}$ such that for every point $(x,y)\in W_1$, for $r<r_2$ we have

$$|\Delta_r(f;x,y)-\varphi_1(x,y)|\leq\frac{r_1}{2},\quad r_2<\frac{r_1}{2}.$$

For the number $\dfrac{r_2}{2}$ there exists a continuous function $F_2(x,y)$ on R_0 such that for all $(x,y)\in R_0$,

$$|F_2(x,y)|<\frac{r_2^2}{2^2},$$

$F_2(x,y)=0$, when $(x,y)\in R_0-\Delta_2$ and almost everywhere on R_0,

$$|\overline{\Delta}(F_1+F_2)-(\varphi_1+\varphi_2)|\leq\frac{r_2^2}{2\cdot 2^2}.$$

Furthermore, there exists a closed set W_2, $|R_0-W_2|<\frac{1}{2^2}$, such that for any point $(x,y)\in W_2$ we have

$$|\Delta_r[(F_1+F_2);x,y]-[\varphi_1(x,y)+\varphi_2(x,y)]|\leq\frac{r_2^2}{2^2}$$

for $r<r_3<\frac{r_2}{2}$.

Let us choose functions $F_1(x,y),\ldots,F_{k-1}(x,y)$. Then for the number $\frac{1}{2^{k-1}}$ there exists a closed set W_{k-1}, $|R_0-W_{k-1}|<\frac{1}{2^{k-1}}$ such that for any point $(x,y)\in W_{k-1}$ and $r<r_k<\frac{r_{k-1}}{2}$ we have

$$\left|\Delta_r\left[\sum_{i=1}^{k-1}(F_i;x,y)\right]-\sum_{i=1}^{k-1}\varphi_i(x,y)\right|<\frac{r_{k-1}^2}{2^{k-1}}. \tag{6.4}$$

Now, by virtue of Lemma 3.6.3, for $\frac{r_k}{2^k}$ there exists a continuous function $F_k(x,y)$ on R_0 such that for all $(x,y)\in R_0$

$$|F_k(x,y)|<\frac{r_k^2}{2^k}, \tag{6.5}$$

$F_k(x,y)=0$ when $(x,y)\overline{\subset}\Delta_k$, and almost everywhere on R_0,

$$|\overline{\Delta}F_k(x,y)-\varphi_k(x,y)|<\frac{r_k^2}{2^{k+1}}.$$

Thus we have obtained the sequences of continuous functions $F_1(x,y),F_2(x,y),\ldots,F_k(x,y)\ldots$; numbers $r_1>r_2>\cdots>r_k>\cdots\to 0$ and sets $W_1,W_2,\ldots,W_k$,

$\dots, |R_0 - W_k| \to 0$ for which the relations (6.4) and (6.5) are fulfilled. Without loss of generality it can be further assumed that

$$r_k < \frac{1}{2}\rho_m, \quad \text{when} \quad v_{m-1} < k \le v_m. \tag{6.6}$$

If the positive integers k and m satisfy the condition $v_{m-1} < k \le v_m$, then we call them the corresponding numbers.

Assume

$$F(x,y) = \sum_{k=1}^{\infty} F_k(x,y).$$

Clearly, the function $F(x,y)$ is continuous on R_0. Let us show that $F_k(x,y)$ is the sought function.

Let E_k be a set of points at which there exists $\overline{\Delta} F_k(x,y)$. It is obvious that, $mE_k = mR_0$.

Assume $E = \bigcap\limits_{k=1}^{\infty} E_k$, then $mE = mR_0$ and on E there exist $\overline{\Delta} F_k(x,y)$, $k = 1, 2, \dots$.

Let $W = \lim\limits_{k\to\infty} W_k$. It can be easily verified that $mW = mR_0$. Introduce the set $\mathbf{E} = E \cap W \cap \Theta$. Thus $m\mathbf{E} = mR_0$.

Let us show that on the set $\mathbf{E}$,

$$\overline{\Delta} F(x,y) = f(x,y). \tag{6.7}$$

Let $\varepsilon > 0$ be a given number, and a point $\mathbf{P}_0(x_0, y_0) \in \mathbf{E}$. We choose the corresponding numbers k_0 and m_0 so that

$$\frac{1}{k_0} < \varepsilon \tag{6.8}$$

and

$$\left| f(x_0,y_0) - \sum_{i=1}^{k-1} \varphi_i(x_0,y_0) \right| < \varepsilon \quad \text{for} \quad k > k_0. \tag{6.9}$$

Further, since $(x_0, y_0) \in \mathbf{E}$, the numbers m_0 and k_0 can be chosen so that the additional conditions

$$(x_0,y_0) \in \Theta_{m_0 - 1}, \quad (x_0,y_0) \in W_k, \quad k \ge k_0.$$

would be fulfilled.

No matter what the number r is, the number k_0 can be chosen so that

$$r_{k+1} < r < r_k \quad \text{for} \quad k > k_0.$$

Let us consider the expression

$$|\Delta_r(F; x_0, y_0) - f(x_0,y_0)| = \left| \Delta_r(F; x_0, y_0) - \sum_{i=1}^{\infty} \varphi_i(x_0,y_0) \right|$$

$$\leq \left| \sum_{i=k}^{\infty} \varphi_i(x_0, y_0) \right| + \left| \Delta_r(F; x_0, y_0) - \sum_{i=1}^{k-1} \varphi_i(x_0, y_0) \right| = I_1 + I_2. \tag{6.10}$$

In view of (6.9),

$$|I_1| < \varepsilon. \tag{6.11}$$

Furthermore,

$$I_2 < \left| \Delta_r \left[\sum_{i=1}^{k-1} (F_i; x_0, y_0) \right] - \sum_{i=1}^{k-1} \varphi_i(x_0, y_0) \right|$$

$$+ |\Delta_r(F_k; x_0, y_0)| + \left| \sum_{i=k+1}^{\infty} \Delta_r(F_i; x_0, y_0) \right| = L_1 + L_2 + L_3. \tag{6.12}$$

Since $k > k_0$, we have $(x_0, y_0) \in W_{k-1}$ and therefore by virtue of (6.4)

$$L_1 < \frac{r_{k-1}^2}{2^{k-1}} < \frac{1}{k_0} < \varepsilon. \tag{6.13}$$

Let k and m_k be the corresponding numbers, i.e.,

$$v_{m_{k-1}} < k \leq v_{m_k}.$$

By condition, $(x_0, y_0) \in \Theta_{m_0 - 1} \subset \Theta_{m_k - 1}$. Therefore for any point (t, τ) from $\Delta_k \subset \mathbf{E}_{m_k}$ we have

$$\rho[(x_0, y_0), (t, \tau)] \geq \rho(\Theta_{m_{k-1}}, \mathbf{E}_{m_k}) = \rho_{m_k}. \tag{6.14}$$

Since $(x_0, y_0) \in \Theta_{m_0 - 1}$, by virtue of (6.14), for any point $(t, \tau) \in \Delta_k$ we have

$$\begin{aligned} \rho[(x_0 + r\cos\varphi, y_0 + r\sin\varphi), (t, \tau)] &\geq \rho[(x_0, y_0), (t, \tau)] \\ -\rho[(x_0 + r\cos\varphi, y_0 + r\sin\varphi), (x_0, y_0)] &\geq \rho_{m_k} - r > \rho_{m_k} - r_k \\ > \rho_{m_k} - \frac{1}{2}\rho_{m_k} &= \frac{1}{2}\rho_{m_k} > 0. \end{aligned}$$

Thus the points (x_0, y_0) and $(x_0 + r\cos\varphi, y_0 + r\sin\varphi)$ lie at a positive distance from Δ_k and therefore $F_k(x_0, y_0) = F_k(x_0 + r\cos\varphi, y_0 + r\sin\varphi) = 0$. Hence

$$L_2 = 0. \tag{6.15}$$

Let us now estimate the last summand:

$$L_3 = \left| \sum_{i=k+1}^{\infty} \Delta_r(F_i; x_0, y_0) \right| = \left| \sum_{i=k+1}^{\infty} \frac{2}{\pi r^3} \int_{C(\mathbf{P}_0; r)} F_i(t, \tau) dS(t, \tau) \right|$$

$$\leq \sum_{i=k+1}^{\infty} \frac{2}{\pi r^3} \int\limits_{C(\mathbf{P}_0;r)} \frac{r_i^2}{2^i} dS(t,\tau) = \sum_{i=k+1}^{\infty} \frac{2r_i^2}{2^i r^3} \cdot 2\pi r$$

$$\leq \frac{8r_{k+1}^2}{2^{k+1}r^2} < \frac{8}{2^{k+1}} < \frac{1}{k_0} < \varepsilon. \tag{6.16}$$

From (6.10),(6.11),(6.12),(6.13),(6.15) and (6.16) it follows that

$$|\Delta_r(F;x_0,y_0) - f(x_0,y_0)| < 3\varepsilon, \quad \text{for} \quad (x_0,y_0) \in \mathbf{E}.$$

Thus $\overline{\Delta}_r F(x_0,y_0) = f(x_0,y_0)$ and therefore

$$\overline{\Delta}_r F(x,y) = f(x,y)$$

almost everywhere.

Theorem 3.6.3 is proved. □

Theorem 3.6.4 ([77], [78] and [80]). *Let $f(x,y)$ be an arbitrary measurable and almost everywhere finite function on S^2. Then there exists a harmonic function $U(\rho,x,y)$ in V^3 such that*

$$\lim_{\rho \to 1} U(\rho,x,y) = f(x,y)$$

almost everywhere on S^2.

Proof. By Theorem 3.6.3, for $f(x,y)$ wee construct a continuous function $F(x,y)$ such that the equality

$$\overline{\Delta} F(x,y) = f(x,y)$$

is fulfilled almost everywhere on S^2.

Consider the expression

$$U(\rho,x,y) = \frac{1}{4\pi} \int\limits_{S^2} D_3 \Big\{ \frac{1-\rho^2}{(1-2\rho\cos\gamma+\rho^2)^{\frac{3}{2}}} \Big\} F(t,\tau) dS^2(t,\tau), \tag{6.17}$$

where D_3 is a Laplace operator on S^2.

It is easy to show that the function $U(\rho,x,y)$ defined by the equality (6.17) is harmonic in V^3. By Theorem 3.4.1, if at the point (x,y) there exists $\overline{\Delta} F(x,y)$, then

$$\lim_{\rho \to 1} U(\rho,x,y) = \overline{\Delta} F(x,y).$$

Since $\overline{\Delta} F(x,y) = f(x,y)$ almost everywhere, we conclude that Theorem 3.6.4 is proved. □

For any measurable boundary function $f(x,y)$, not necessarily finite almost everywhere, O.P. Dzagnidze investigated radial boundary values by quite a different method ([17], [18] and [19]). I.I. Privalov ([51], [52]) and E.D. Solomentsev ([60]–[64]) studied the characteristic properties of harmonic functions representable by Green–Stieltjes and Green–Labesgue type integrals in general domains.

3.7 Representation by the Laplace Series of an Arbitrary Measurable Function Defined on the Unit Sphere S^2

In 1915, N.N. Luzin formulated the following problem:

Let $f(x)$ be an arbitrary measurable function* defined on $[0; 2\pi]$. Does there exist a trigonometric series

$$\frac{a_0}{2} + \sum_{n=1}^{\infty}(a_n \cos nx + b_n \sin nx), \tag{7.1}$$

which is convergent or summable some method to a function $f(x)$ almost everywhere on $[0; 2\pi]$?

N.N. Luzin ([43], [44]) proved that if $f(x)$ is almost everywhere a finite measurable function, then there exists a trigonometric series of form (7.1.) which is almost everywhere summable to $f(x)$ by both the Abel method and the Riemann method.

In 1925, I.I. Privalov and N.N. Luzin ([53], p. 309) proved that there exists a trigonometric series which is summable to $+\infty$ almost everywhere by the Abel method (with respect to the normal).

Yu.B. Hermyyer ([8]) proved that there exists no trigonometric series summable by the Riemann method to $+\infty$ on a set of positive measure.

The fundamental result in this direction was obtained by D.E. Menshof ([48]), (see also [1]) in 1941. He established that for any almost everywhere finite on $[0, 2\pi]$ measurable function there exists a trigonometric series of form (7.1) which converges to that function almost everywhere on $[0; 2\pi]$.

In 1950, D.E. Menshof ([49]) proved that for a measurable function that may transform to $+\infty$ or $-\infty$ on a set of positive measure, there exists a trigonometric series converging to that function in measure.

In 1988, S.V. Konuagin [40] proved that a trigonometric series is not convergent to $+\infty$ on a set of positive measure. According to Menshof and Konyagin, for a function f to be representable by an almost everywhere converging trigonometric series, it is necessary and sufficient that this function be measurable and almost everywhere finite on $[0; 2\pi]$.

Problems of representation of measurable functions of one variable by series with respect to various systems of functions are studied with sufficient thoroughness. The modern state of this issue is discussed in detail in [29], [31], [32], [50], [70], [71], [113] and [114].

In this section we consider the problem of representation of a measurable and almost everywhere finite function defined on the unit sphere S^2 by a Laplace series, namely, we prove the theorem which is an analogue of the above-mentioned Luzin's theorem.

---

*$f(x)$ may be equal to $+\infty$ or $-\infty$ on sets of positive measure

Consider the Laplace series (see (1.12))

$$\sum_{m=0}^{\infty} Y_m(\theta,\varphi), \tag{7.2}$$

where $Y_m(\theta,\varphi)$ is a spherical harmonic of order m $(0\le\theta\le\pi; -\pi\le\varphi\le\pi)$.

Of the terms of the series (7.2) we compose the series

$$\sum_{m=1}^{\infty} \Omega Y_m(\theta,\varphi) = -\sum_{m=1}^{\infty} \frac{Y_m(\theta,\varphi)}{m(m+1)}, \tag{7.3}$$

where the operator Ω is defined by the equality ([56], p. 293)

$$\Omega f(x) = \int_{S^2} f(y)G(x,y)dS^2(y),$$

$$G(x,y) = \frac{1}{2\pi}\ln\sin\frac{\overset{\wedge}{xy}}{2}.$$

A first order generalized Laplace operator of a function $f(x)=\varphi(\theta,\varphi)$ $(0\le\theta\le\pi;\ -\pi\le\varphi\le\pi)$ at a point $x\in S^2$ denoted by $\overline{\Delta}f(x)$ (see Section 3.3), is defined by the equality ([56], p. 288))

$$\overline{\Delta}f(x) = \lim_{h\to 0}\frac{\frac{1}{2\pi\sinh}\int_{(x,t)=\cosh} f(t)dS(t) - f(x))}{\sin^2\frac{h}{2}}.$$

Let us assume that (7.3) is the Fourier–Laplace series of the function $F(x) = F(\theta,\varphi)\in L(S^2)$.

The series (7.2) is called summable by the Riemann method (or, shortly, R-summable) to $I_0(\theta,\varphi)+S(\theta,\varphi)$ at a point $x(\theta,\varphi)$ if ([56], p. 289)

$$\overline{\Delta}F(\theta,\varphi) = S(\theta,\varphi).$$

A point $x\in S^2$ is called an L-point of the function f if

$$\lim_{h\to 0}\frac{1}{h^2}\int_{D^2(x;h)} |f(y)-f(x)|dS^2(y) = 0.$$

From [56] (p. 296) we have the following

Theorem A. *If $f\in L(S^2)$, then its Fourier–Laplace series is summable by the R method at all L-points of that function to $f(x)$.*

The following theorem is valid [92]).

Theorem 3.7.1. *If $f(x)$ is a finite measurable function almost everywhere on S^2, then there exists a Laplace series which everywhere on S^2 is summable to $f(x)$ by both the Abel method and the R method.*

Proof. Let $f(x)$ be a measurable function on S^2, finite almost everywhere. By Theorem 3.6.3, there exists a continuous function $F(x)$ such that $\overline{\Delta}F(x) = f(x)$ almost everywhere on S^2. Let (7.2) be a Fourier–Laplace series of the function $F(x)$. Differentiating the series (7.2) termwise, we obtain the Laplace series

$$\sum_{m=0}^{\infty} D_3 Y_m(\theta, \varphi) = -\sum_{m=1}^{\infty} m(m+1) Y_m(\theta, \varphi), \tag{7.4}$$

where D_3 is a Laplace operator on S^2 (see Section 3.1).

By Theorem 3.4.1, the series (7.4) is summable almost everywhere on $S^@$ by the Abel method to $f(x)$.

Furthermore, at all points $x \in S^2$ at which there exists a finite $\overline{\Delta}F(x)$, the series (7.4) is summable by the R method at a point x to $\overline{\Delta}F(x)$, and $\overline{\Delta}F(x)$=f(x) almost everywhere.

The theorem is proved. □

Remark. We do not know the following: 1) whether the theorem proved above remains valid when $f(x) = +\infty$ or $f(x) = -\infty$ on a set of positive measure; 2) whether analogous statements are valid for the method (C, α), $\alpha \geq 0$, and whether α depends on space dimension.

Chapter 4

Boundary Properties of Derivatives of the Poisson Integral for a Space R_+^{k+1} $(k > 1)$

4.1 Generalized Partial First Order Derivatives of a Function of Several Variables

R^k is the k-dimensional Euclidean space ($R = R^1$).

e_i ($i = 1, 2, \ldots, k$) is the coordinate vector.

Let (see [31], p. 174) $M = \{1, 2, \ldots, k\}$, ($k \in N$, $k \geq 2$) and B be an arbitrary subset of the set M, $B' = M|_B$ be a complement of the set B with respect to M. For $x \in R^k$ and $B \subset M$, by $x_{_B}$ we denote a point of R^k whose coordinates with indices from the set B coincide with the corresponding coordinates of the point x, and the coordinates with indices from the set B' are zeros ($x_{_M} = x$, $B|_i = B|_{\{i\}}$). If $B = \{i_1, i_2, \ldots, i_s\}$ $1 \leq s \leq k$ ($i_l < i_r$ for $l < r$), then $\overline{x}_{_B} = (x_{i_1}, x_{i_2}, \ldots, x_{i_s}) \in R^s$; $m(B)$ is a number of elements of the set B; $\widetilde{L}(R^k)$ is a set of functions $f(x) = f(x_1, x_2, \ldots, x_k)$, such that

$$\frac{f(x)}{(1 + |x|^2)^{\frac{k+1}{2}}} \in L(R^k).$$

Let $u \in R$. For the function $f(x)$ we consider the following derivatives.

1. We denote the limit

$$\lim_{(u, \overline{x}_{_B}) \to (0, \overline{x}^0_{_B})} \frac{f(x_{_B} + x^0_{_{B'}} + ue_i) - f(x_{_B} + x^0_{_{B'}})}{u}$$

a) by $f'_{x_i}(x^0)$ if $B = \varnothing$,

b) by $\mathbf{D}_{x_i(\overline{x}_B)}f(x^0)$ if $i \in B'$,

c) by $\overline{\mathbf{D}}_{x_i(\overline{x}_B)}f(x^0)$ if $i \in B$.

2. The limit

$$\lim_{(u,\overline{x}_B)\to(0,\overline{x}^0_B)} \frac{f(x_B + x^0_{B'} + ue_i) - f(x_B + x^0_{B'} - ue_i)}{2u}$$

is denoted

a) by $\mathbf{D}_{x_i}{}^* f(x^0)$ if $B = \varnothing$,

b) by $\mathbf{D}^*_{x_i}(\overline{x}_B)f(x^0)$ if $i \in B'$,

c) by $\overline{\mathbf{D}}^*_{x_i(\overline{x}_B)}f(x^0)$ if $i \in B$.

The following statements are valid:

1) If $B_2 \subset B_1$, then the existence of $\mathbf{D}_{x_i}(\overline{x}_{B_1})f(x^0)$ implies the existence of $\mathbf{D}_{x_i}(\overline{x}_{B_2})f(x^0)$, and $\mathbf{D}_{x_i}(\overline{x}_{B_1})f(x^0) = f'_{x_i}(x^0) = \mathbf{D}_{x_i}(\overline{x}_{B_2})f(x^0)$. The converse is not true.

2) The existence of $\overline{\mathbf{D}}_{x_i}(\overline{x}_{B_1})f(x^0)$ implies the existence of $\overline{\mathbf{D}}_{x_i}(\overline{x}_{B_2})f(x^0)$ and their equivalence.

3) The existence of $\overline{\mathbf{D}}_{x_i}(\overline{x}_B)f(x^0)$ implies the existence of $\overline{\mathbf{D}}_{x_i}(\overline{x}_{B|i})f(x^0)$, and $\overline{\mathbf{D}}_{x_i(\overline{x}_B)}f(x^0) = \overline{\mathbf{D}}_{x_i}(\overline{x}_{B|i})f(x^0) = f'_{x_i}(x^0)$.

4) If $f'_{x_i}(x)$ is continuous at the point x^0, then for any $B \subset M$ all derivatives $\overline{\mathbf{D}}_{x_i}(\overline{x}_B)f(x^0)$ exist, and

$$\overline{\mathbf{D}}_{x_i}(\overline{x}_B)f(x^0) = f'_{x_i}(x^0).$$

Indeed, by the Lagrange theorem we have

$$\begin{aligned}\frac{f(x_B + x^0_{B'} + ue_i) - f(x_B + x^0_{B'})}{u} &= \frac{f'_{x_i}[x_B + x^0_{B'} + \theta(x)ue_i]u}{u} \\ &= f'_{x_i}[x_B + x^0_{B'} + \theta(x)ue_i], \quad 0 < \theta < 1,\end{aligned}$$

from which it follows that Statement 4) is valid.

5) There exists a function $f(x)$ for which $\overline{\mathbf{D}}_{x_i(x)}f(x^0)$ exist, but, on an everywhere dense set, $f'_{x_i}(x)$ do not exist (see Section 1.1) in the neighborhood of the point x^0 (see Section 1.1).

6) If $f(x)$ has, at the point x^0, the finite derivatives

$$\mathbf{D}_{x_1(x_2,x_3,\dots,x_k)}f(x^0), \mathbf{D}_{x_2(x_3,\dots,x_k)}f(x^0), \dots, \mathbf{D}_{x_{k-1}(x_k)}f(x^0),$$

then the continuity of f at the point x^0 in the argument x_k is the necessary and sufficient condition for the function $f(x)$ to be continuous at the point x^0 (see [26], p. 15).

7) The existence of the derivatives $\mathbf{D}_{x_1(x_2,x_3,\dots,x_k)}f(x^0), \mathbf{D}_{x_2(x_3,\dots,x_k)}f(x^0), \dots,$ $\mathbf{D}_{x_{k-1}(x_k)}f(x^0)$ and $f'_{x_k}(x^0)$ implies the existence of the differential $df(x^0)$ (see [26], p. 16).

In the sequel, it will be assumed that $f \in \widetilde{L}(R^k)$.

4.2 The Boundary Properties of First Order Partial Derivatives of the Poisson Integral for a Half-Space R_+^{k+1} $(k > 1)$

The integral

$$U(f; x, x_{k+1}) = \frac{x_{k+1}\Gamma\left(\frac{k+1}{2}\right)}{\pi^{\frac{k+1}{2}}} \int\limits_{R^k} \frac{f(t)dt}{(|t-x|^2 + x_{k+1}^2)^{\frac{k+1}{2}}}$$

$$= \frac{\Gamma\left(\frac{k+1}{2}\right)}{\pi^{\frac{k+1}{2}}} \int\limits_{R^k} P(t-x, x_{k+1}) f(t)dt,$$

where $P(t-x, x_{k+1}) = \dfrac{x_{k+1}}{(|t-x|^2 + x_{k+1}^2)^{\frac{k+1}{2}}}$, $x, t \in R^k$ is the kernel, is called the Poisson integral for a half-space $R_+^{k+1} = \{(x, x_{k+1}) \in R^{k+1} : x \in R^k, x_{k+1>0}\}$.

It is shown in [93] (p. 25) that there exists a continuous function of two variables $f(x, y) \in L(R^2)$ such that at some point (x_0, y_0) there exist finite partial derivatives $f'_x(x_0, y_0)$ and $f'_y(x_0, y_0)$, but for this function the integrals $\dfrac{\partial U(f; x, y, z)}{\partial x}$ and $\dfrac{\partial U(f; x, y, z)}{\partial y}$ ($U(f; x, y, z)$ is the Poisson integral for R_+^3) have no boundary values at the point (x_0, y_0) even with respect to the norm.

Hence there naturally arises the question whether it is possible to generalize the notion of derivatives of a function of several variables so that a Fatou type theorem be true for the integral $U(f; x, x_{k+1})$.

In this section, for the derivatives introduced in Section 4.1, we prove the Fatou type theorems on the boundary behavior of first order partial derivatives of the Poisson integral for the half-space R_+^{k+1} ([83], [96]–[104]). In particular, it will be shown that the boundary properties of derivatives of the Poisson integral for a half-space depend essentially on how the Poisson integral density is differentiable. We construct the examples illustrating that the obtained results are unimprovable (in a certain sense).

When investigating the boundary properties of partial derivatives $\dfrac{\partial}{\partial \theta} U_f(r, \theta, \varphi)$ and $\dfrac{\partial}{\partial \varphi} U_f(r, \theta, \varphi)$ of the spherical Poisson integral $U_f(r, \theta, \varphi)$ for a summable function $f(\theta, \varphi)$ on a rectangle $[0, \pi] \times [0, 2\pi]$, O.P. Dzagnidze introduced the notion of a two-sided angular limit ([23], p. 63) which extends to R_+^{k+1} as follows: if the point $N \in R_+^{k+1}$ tends to the point $\mathbf{P}(x^0, 0)$ provide that

$$\frac{x_{k+1}}{\sqrt{\sum\limits_{i\in B}(x_i-x_i^0)^2}}\geq C>0,^*$$

then we write $N(x,x_{k+1})\xrightarrow[x_B]{\wedge}\mathbf{P}(x^0,0)$.

If $B=M$, then we have an angular tendency and we write $N(x,x_{k+1})\xrightarrow{\wedge}\mathbf{P}(x^0,0)$. Finally, the notation $N(x,x_{k+1})\longrightarrow\mathbf{P}(x^0,0)$ means that the point $N(x,x_{k+1})$ tends to $\mathbf{P}(x^0,0)$ without any restrictions on the tendency and remains in R_+^{k+1}.

Lemma 4.2.1. *For any $(x,x_{k+1})\in R_+^{k+1}$ and $i=\overline{1,k}$ the following statements are valid:*

1) $I_1=\int\limits_{R^k}\frac{(t_i-x_i)f(t-t_ie_i)dt}{\left(|t-x|^2+x_{k+1}^2\right)^{\frac{k+3}{2}}}=0$;

2) $I_2=\frac{(k+1)x_{k+1}\Gamma\left(\frac{k+1}{2}\right)}{\pi^{\frac{k+1}{2}}}\int\limits_{R^k}\frac{(t_i-x_i)^2dt}{\left(|t-x|^2+x_{k+1}^2\right)^{\frac{k+3}{2}}}=1$;

3) $\lim\limits_{x_{k+1}\to0}\sup\limits_{|t|\geq\delta>0}\left|\frac{\partial P(t,x_{k+1})}{\partial t_i}\right||t|=0$;

4) $\int\limits_{R^k}\left|\frac{\partial P(t,x_{k+1})}{\partial t_i}\right||t|dt=O(1)$;

5) $\int\limits_{R^k}\left|\frac{\partial P(t-x,x_{k+1})}{\partial t_i}\right||t|dt=O(1)$ *for* $\frac{x_{k+1}}{|x_i|}\geq C>0$.

Proof. We have

$$I_1=\int\limits_{R^k}\frac{(t_i-x_i)f(t-t_ie_i)}{\left(|t-x|^2+x_{k+1}^2\right)^{\frac{k+3}{2}}}dt$$

$$=\int\limits_{R^{k-1}}f(x+t-t_ie_i)dS(\bar{t}_{M|i})\int\limits_{-\infty}^{\infty}\frac{t_idt_i}{\left(|t|^2+x_{k+1}^2\right)^{\frac{k+3}{2}}}=0.$$

Next,

$$I_2=\frac{(k+1)x_{k+1}\Gamma\left(\frac{k+1}{2}\right)}{\pi^{\frac{k+1}{2}}}\int\limits_{R^k}\frac{t_i^2dt}{\left(|t|^2+x_{k+1}^2\right)^{\frac{k+3}{2}}}.$$

Passing to the spherical coordinates $(\rho,\theta_1,\theta_2,\dots,\theta_{k-2},\varphi)$, we have

$$\int\limits_{R^k}\frac{t_i^2dt}{\left(|t|^2+x_{k+1}^2\right)^{\frac{k+3}{2}}}=\int\limits_{R^k}\frac{\rho^2\sin^2\theta_1\sin^2\theta_2\dots\sin^2\theta_{i-1}\cos^2\theta_i}{\left(\rho^2+x_{k+1}^2\right)^{\frac{k+3}{2}}}$$

*Here and everywhere below, by C we denote absolute positive constants which may, generally speaking, be different in various relations.

$$\times \rho^{k-1} \sin^{k-2}\theta_1 \dots \sin^{k-i}\theta_{i-1} \dots \sin\theta_{k-2} d\rho d\theta_1 \dots d\theta_{k-2} d\varphi$$

$$= \frac{2\pi}{x_{k+1}} \int_0^\infty \frac{\rho^{k+1} d\rho}{(1+\rho^2)^{\frac{k+3}{2}}} \overbrace{\int_0^\pi \dots \int_0^\pi}^{k-2} \sin^k\theta_1 \sin^{k-1}\theta_2 \dots \sin^{k-i+2}\theta_{i-1}$$
$$\times \cos^2\theta_i \sin^{k-i-1}\theta_i \dots \sin\theta_{k-2} d\theta_1 d\theta_2 \dots d\theta_{k-2}.$$

Using the well-known equality ([4], p. 383)

$$\int_0^\pi \sin^{\mu-1}\theta d\theta = \frac{\Gamma\left(\frac{\mu}{2}\right)}{\Gamma\left(\frac{\mu+1}{2}\right)}\sqrt{\pi},$$

we can show that

$$\overbrace{\int_0^\pi \dots \int_0^\pi}^{k-2} \sin^k\theta_1 \sin^{k-1}\theta_2 \dots \sin^{k-i+2}\theta_{i-1} \cos^2\theta_i \sin^{k-i-1}\theta_i$$
$$\times \sin^{k-i-2}\theta_{i+1} \dots \sin\theta_{k-2} d\theta_1 d\theta_2 \dots d\theta_{k-2} = \frac{\pi^{\frac{k-2}{2}}}{k\Gamma\left(\frac{k}{2}\right)}.$$

On the other hand, ([10], p. 311, No.10, and p. 1023, No 8.756), it can be verified that

$$\int_0^\infty \frac{\rho^{k+1} d\rho}{(1+\rho^2)^{\frac{k+3}{2}}} = \frac{k\Gamma\left(\frac{k}{2}\right)\sqrt{\pi}}{2(k+1)\Gamma\left(\frac{k+1}{2}\right)}.$$

As a result, we have

$$I_2 = \frac{(k+1)x_{k+1}\Gamma\left(\frac{k+1}{2}\right)}{\pi^{\frac{k+1}{2}}} \cdot \frac{2\pi}{x_{k+1}} \cdot \frac{k\Gamma\left(\frac{k}{2}\right)\sqrt{\pi}}{2(k+1)\Gamma\left(\frac{k+1}{2}\right)} \cdot \frac{\pi^{\frac{k-2}{2}}}{k\Gamma\left(\frac{k}{2}\right)} = 1.$$

The validity of Statements 3), 4) and 5) is proved analogously.
Thus Lemma 4.2.1 is proved.
The following theorem is valid. □

Theorem 4.2.1. (a) *If at the point x^0 there exists a finite derivative $\overline{\mathbf{D}}_{x_i(x)} f(x^0)$, $1 \le i \le k$, then*

$$\lim_{(x,x_{k+1})\to(x^0,0)} \frac{\partial U(f; x, x_{k+1})}{\partial x_i} = \frac{\partial f(x^0)}{\partial x_i}. \tag{2.1}$$

(b) *There exists a continuous function* $g \in L(R^k)$ *such that at the point* $x^0 = (0,0,\dots,0) = 0$ *the equality* $\overline{\mathbf{D}}_{x_i(x_B)} f(0) = 0$, $i = \overline{1,k}$, *holds for any* $B \subset M$ *with the property* $m(B) < k$, *but the limit*

$$\lim_{x_{k+1}\to 0+} \frac{\partial U(g;0,x_{k+1})}{\partial x_i}, \quad i = \overline{1,k}$$

do not exist.

Proof of Item (a). Let $x^0 = 0$, $C_k = \frac{(k+1)\Gamma\left(\frac{k+1}{2}\right)}{\pi^{\frac{k+1}{2}}}$. It is not difficult to verify that

$$\frac{\partial U(f;x,x_{k+1})}{\partial x_i} = C_k x_{k+1} \int\limits_{R^k} \frac{(t_i - x_i) f(t)dt}{\left(|t-x|^2 + x_{k+1}^2\right)^{\frac{k+3}{2}}}$$
$$= C_k x_{k+1} \int\limits_{R^k} \frac{t_i f(x+t)dt}{\left(|t|^2 + x_{k+1}^2\right)^{\frac{k+3}{2}}}.$$

In view of Lemma 4.2.1, we have

$$\frac{\partial U(f;x,x_{k+1})}{\partial x_i} - \overline{\mathbf{D}}_{x_i(x)} f(o) = C_k x_{k+1} \int\limits_{R^k} \frac{t_i^2}{\left(|t|^2 + x_{k+1}^2\right)^{\frac{k+3}{2}}}$$
$$\times \Big[\frac{f(x+t) - f(x+t-t_ie_i)}{t_i} - \overline{\mathbf{D}}_{x_i(x)} f(o)\Big] dt = I_1 + I_2,$$

where

$$I_1 = C_k x_{k+1} \int\limits_{V_\delta}, \quad I_2 = C_k x_{k+1} \int\limits_{CV_\delta};$$

V_δ is a ball with center at the point o, of radius δ. Let $\varepsilon > 0$. We choose $\delta > 0$ such that

$$\Big|\frac{f(x+t) - f(x+t-t_ie_i)}{t_i} - \overline{\mathbf{D}}_{x_i(x)} f(0)\Big| < \varepsilon$$

for $|x| < \delta$ and $|t| < 2\delta$.

Thus we have

$$|I_1| < C_k x_{k+1} \varepsilon \int\limits_{V_\delta} \frac{t_i^2 dt}{\left(|t|^2 + x_{k+1}^2\right)^{\frac{k+3}{2}}} < C_k x_{k+1} \varepsilon \int\limits_{R^k} \frac{t_i^2 dt}{\left(|t|^2 + x_{k+1}^2\right)^{\frac{k+3}{2}}} = \varepsilon. \tag{2.2}$$

It is also easy to show that

$$\lim_{(x,x_{k+1})\to(x^0,0)} I_2 = 0. \tag{2.3}$$

(2.2) and (2.3) show that the equality (2.1) is valid.

Proof of Item (b). Let $\mathbf{D} = (0 \le t_1 < \infty; 0 \le t_2 < \infty, \dots, 0 \le t_k < \infty)$. We find the function g as follows:

$$g(t) = \begin{cases} \sqrt[k+1]{t_1 t_2, \dots t_k}, & \text{if} \quad (t_1, t_2, \dots, t_k) \in D, \\ 0, & \text{if} \quad (t_1, t_2, \dots, t_k) \in CD. \end{cases}$$

We can see that $g(t)$ is continuous in R^k and $\overline{\mathbf{D}}_{x_i(\overline{x}_B)} f(o) = 0$, $i = \overline{1,k}$ for any B when $m(B) < k$.

If in the integral

$$\frac{\partial U(g; x, x_{k+1})}{\partial x_i} = C_k x_{k+1} \int\limits_{R^k} \frac{(t_i - x_i) g(t) dt}{(|t - x|^2 + x_{k+1}^2)^{\frac{k+3}{2}}}$$

we pass to the spherical coordinates, then for the considered function we have

$$\frac{\partial U(g; 0, x_{k+1})}{\partial x_i} = C_k x_{k+1} \int\limits_{R^k} \frac{t_i g(t) dt}{(|t|^2 + x_{k+1}^2)^{\frac{k+3}{2}}}$$

$$= C x_{k+1} \int\limits_0^\infty \frac{\rho \sqrt[k+1]{\rho^k}}{(\rho^2 + x_{k+1}^2)^{\frac{k+3}{2}}} \rho^{k-1} d\rho$$

$$= C x_{k+1} \int\limits_0^\infty \frac{\rho^{k+\frac{k}{k+1}} d\rho}{(\rho^2 + x_{k+1}^2)^{\frac{k+3}{2}}} > C x_{k+1} \int\limits_0^{x_{k+1}} \frac{\rho^{k+\frac{k}{k+1}} d\rho}{(\rho^2 + x_{k+1}^2)^{\frac{k+3}{2}}}$$

$$> C x_{k+1} \int\limits_0^{x_{k+1}} \frac{\rho^{k+\frac{k}{k+1}} d\rho}{x_{k+1}^{k+3}} = \frac{C}{\sqrt[k+1]{x_{k+1}}},$$

whence

$$\lim_{x_{k+1} \to 0+} \frac{\partial U(g; 0, x_{k+1})}{\partial x_i} = +\infty.$$

The theorem is proved. □

Corollary 4.2.1. *If at the point x^0 there exist finite derivatives $\overline{\mathbf{D}}_{x_i(x)} f(x^0)$, $i = \overline{1,k}$, then*

$$\lim_{(x, x_{k+1}) \to (x^0, 0)} d_x U(f; x, x_{k+1}) = df(x^0).$$

Corollary 4.2.2. *If f has a continuous partial derivative $f'_{x_i}(x)$ at the point x^0, then*

$$\lim_{(x, x_{k+1}) \to (x^0, 0)} \frac{\partial U(f; x, x_{k+1})}{\partial x_i} = f'_{x_i}(x^0).$$

Corollary 4.2.3. (a) *If f is continuously differentiable at the point x^0, then*

$$\lim_{(x,x_{k+1})\to(x^0,0)} d_xU(f;x,x_{k+1}) = df(x^0).$$

(b) *There exists a differentiable function $g(t_1,t_2)$ at the point $(0,0)$ such that $dg(0,0)=0$, but the limits*

$$\lim_{(x_1,x_2,x_3)\to(0,0,0)} \frac{\partial U(g;x_1,x_2,x_3)}{\partial x_1}; \qquad \lim_{(x_1,x_2,x_3)\to(0,0,0)} \frac{\partial U(g;x_1,x_2,x_3)}{\partial x_2}$$

do not exist.

Proof of Item (b). Assume $\mathbf{D}=[0,1;0,1]$. Let

$$g(t_1,t_2)=\begin{cases}\sqrt[5]{t_1^3t_2^3} & \text{when } (t_1,t_2)\in D,\\ 0 & \text{when } (t_1,t_2)\in]-\infty,0;0,\infty[\cup]-\infty,\infty;-\infty,0],\end{cases}$$

and on the set $]0,\infty;0,\infty[\backslash D$ we extend it continuously keeping in mind that the condition $g\in L(R^2)$ is fulfilled. It is easy to verify that $g(t_1,t_2)$ is differentiable at the point $(0,0)$, and

$$g'_{t_1}(0,0)=g'_{t_2}(0,0)=0.$$

Let $(x_1,x_2,x_3)\to(0,0,0)$ for $x_1=0$, $x_3=x_2^2$, $x_2>0$. Then for the considered function

$$\frac{\partial U(g;0,x_2,x_3)}{\partial x_1}=\frac{3x_3}{2\pi}\int_0^\infty\int_0^\infty \frac{t_1g(t_1,t_2)dt_1dt_2}{[t_1^2+(t_2-x_2)^2+x_2^4]^{\frac52}}$$

$$=\frac{3x_2^2}{2\pi}\int_0^\infty\int_{-x_2}^\infty \frac{t_1g(t_1,t_2+x_2)dt_1dt_2}{(t_1^2+t_2^2+x_2^4)^{\frac52}}=\frac{3x_2^2}{2\pi}\int_0^\infty\int_{-x_2}^\infty \frac{t_1\sqrt[5]{t_1^3(t_2+x_2)^3}dt_1dt_2}{(t_1^2+t_2^2+x_2^4)^{\frac52}}$$

$$>\frac{3x_2^2}{2\pi}\int_{x_2^2}^{2x_2^2}\int_{x_2^2}^{2x_2^2} \frac{t_1\sqrt[5]{t_1^3}\sqrt[5]{x_2^3}dt_1dt_2}{(t_1^2+t_2^2+x_2^4)^{\frac52}}>\frac{3x_2^2x_2^{\frac35}}{2\pi}\int_{x_2^2}^{2x_2^2}\int_{x_2^2}^{2x_2^2} \frac{x_2^2\cdot x_2^{\frac65}dt_1dt_2}{(4x_2^4+4x_2^4+x_2^4)^{\frac52}}$$

$$=\frac{1}{162\pi\sqrt[5]{x_2}}\to\infty \quad \text{as} \quad x_2\to 0+.$$

The theorem is proved. □

Theorem 4.2.2. *If at the point x^0 there exists a finite derivative $\mathbf{D}_{x_i(\overline{x}_M|i)}f(x^0)$, then*

$$\lim_{(x,x_{k+1})\underset{x_i}{\xrightarrow{\wedge}}(x^0,0)} \frac{U(f;x,x_{k+1})}{\partial x_i}=\frac{\partial f(x^0)}{\partial x_i}.$$

Proof. Let $x^0 = 0$. By Lemma 4.2.1, we have the equality

$$\frac{U(f;x,x_{k+1})}{\partial x_i} - \mathbf{D}_{x_i(\overline{x}_{M|i})} f(0) = C_k x_{k+1} \int\limits_{R^k} \frac{t_i(t_i - x_i)}{\left(|t-x|^2 + x_{k+1}^2\right)^{\frac{k+3}{2}}}$$
$$\times \left[\frac{f(t) - f(t - t_i e_i)}{t_i} - \mathbf{D}_{x_i(\overline{x}_{M|i})} f(0)\right] dt = I_1 + I_2,$$

where $I_1 = C_k x_{k+1} \int\limits_{V_\delta}$, $I_2 = C_k x_{k+1} \int\limits_{CV_\delta}$.

Let $\varepsilon > 0$, and we choose $\delta > 0$ such that the inequality

$$\left|\frac{f(t) - f(t - t_i e_i)}{t_i} - \mathbf{D}_{x_i(\overline{x}_{M|i})} f(0)\right| < \varepsilon$$

is fulfilled for $|t| < \delta$.

By virtue of the above reasoning,

$$|I_1| < C_k x_{k+1} \varepsilon \int\limits_{R^k} \frac{|t_i(t_i - x_i)| dt}{\left(|t-x|^2 + x_{k+1}^2\right)^{\frac{k+3}{2}}} < C_k x_{k+1} \varepsilon \int\limits_{R^k} \frac{t_i^2 dt}{\left(|t|^2 + x_{k+1}^2\right)^{\frac{k+3}{2}}}$$
$$+ C_k x_{k+1} \varepsilon |x_i| \int\limits_{R^k} \frac{|t_i| dt}{\left(|t|^2 + x_{k+1}^2\right)^{\frac{k+3}{2}}} = \varepsilon + C x_{k+1} \varepsilon |x_i| \int\limits_0^\infty \frac{\rho^k d\rho}{\left(\rho^2 + x_{k+1}^2\right)^{\frac{k+3}{2}}}$$
$$= \varepsilon + \frac{C x_{k+1}^{k+2} |x_i| \varepsilon}{x_{k+1}^{k+3}} \int\limits_0^\infty \frac{\rho^k d\rho}{\left(1 + \rho^2\right)^{\frac{k+3}{2}}} = \left(1 + \frac{C|x_i|}{x_{k+1}}\right)\varepsilon.$$

whence we have

$$\lim_{(x,x_{k+1}) \xrightarrow[x_i]{\wedge} (x^0,0)} I_1 = 0.$$

Analogously, we can prove that

$$\lim_{(x,x_{k+1}) \xrightarrow[x_i]{\wedge} (x^0,0)} I_2 = 0.$$

The theorem is proved. □

Theorem 4.2.3. *If at the point x^0 there exist finite derivatives*

$$\mathbf{D}_{x_i(\overline{x}_{M|i})} f(x^0) \quad \text{and} \quad \mathbf{D}_{x_j(\overline{x}_B)} f(x^0), \quad i \neq j, \quad B = M|\{i,j\},$$

then

$$\lim_{(x,x_{k+1}) \xrightarrow[x_i]{\wedge} (x^0,0)} \frac{\partial U(f;x,x_{k+1})}{\partial x_i} = \frac{\partial f(x^0)}{\partial x_i},$$

$$\lim_{(x,x_{k+1})\xrightarrow[x_ix_j]{\wedge}(x^0,0)} \frac{\partial U(f;x,x_{k+1})}{\partial x_j} = \frac{\partial f(x^0)}{\partial x_j}.$$

Proof. Let $x^0 = 0$. By Lemma 4.2.1, we have

$$\frac{\partial U(f;x,x_{k+1})}{\partial x_j} = C_k x_{k+1} \int\limits_{R^k} \frac{(t_j - x_j)f(t)dt}{(|t-x|^2 + x_{k+1}^2)^{\frac{k+3}{2}}}$$

$$= C_k x_{k+1} \int\limits_{R^k} \frac{(t_j - x_j)\{[f(t) - f(t - t_ie_j)] + [f(t - t_ie_i) - f(t - t_ie_i - t_je_j)]\}dt}{(|t-x|^2 + x_{k+1}^2)^{\frac{k+3}{2}}}$$

$$= C_k x_{x+1} \int\limits_{R^k} \frac{(t_j - x_j)[f(t) - f(t - t_ie_i)]}{(|t-x|^2 + x_{k+1}^2)^{\frac{k+3}{2}}} dt$$

$$+ C_k x_{k+1} \int\limits_{R^k} \frac{(t_j - x_j)[f(t - t_ie_i) - f(t - t_ie_i - t_je_j)]}{(|t-x|^2 + x_{k+1}^2)^{\frac{k+3}{2}}} dt = I_1 + I_2,$$

where

$$I_1 = C_k x_{k+1} \int\limits_{R^k} \frac{t_i(t_j - x_j)}{(|t-x|^2 + x_{k+1}^2)^{\frac{k+3}{2}}} \cdot \frac{f(t) - f(t - t_ie_i)}{t_i} dt$$

$$= C_k x_{k+1} \int\limits_{R^k} \frac{t_i(t_j - x_j)}{(|t-x|^2 + x_{k+1}^2)^{\frac{k+3}{2}}} \Big[\frac{f(t) - f(t - t_ie_i)}{t_i} - \mathbf{D}_{x_i(\overline{x}_{M|i})} f(0)\Big] dt$$

$$+ C_k x_{k+1} \mathbf{D}_{x_i(\overline{x}_{M|i})} f(0) \int\limits_{R^k} \frac{t_i(t_j - x_j)dt}{(|t-x|^2 + x_{k+1}^2)^{\frac{k+3}{2}}} = I_1' + I_1''.$$

It is easy to see that $I_1'' = 0$ and

$$|I_1'| < C_k x_{k+1} \int\limits_{R^k} \frac{|t_i(t_j - x_j)|}{(|t-x|^2 + x_{k+1}^2)^{\frac{k+3}{2}}} \Big|\frac{f(t) - f(t - t_ie_i)}{t_i} - \mathbf{D}_{x_i(\overline{x}_{M|i})} f(0)\Big| dt.$$

Reasoning analogously as in proving Theorem 4.2.2, we obtain

$$\lim_{(x,x_{k+1})\xrightarrow[x_i]{\wedge}0} I_1' = \lim_{(x,x_{k+1})\xrightarrow[x_i]{\wedge}0} I_1 = 0.$$

Let us now show that

$$\lim_{(x,x_{k+1})\xrightarrow[x_j]{\wedge}0} I_2 = \mathbf{D}_{x_j(\overline{x}_{M|\{i,j\}})} f(0).$$

Indeed,

$$I_2 = C_k x_{k+1} \int\limits_{R^k} \frac{|t_j(t_j - x_j)|}{(|t-x|^2 + x_{k+1}^2)^{\frac{k+3}{2}}}$$
$$\times\Big[\frac{f(t-t_ie_i) - f(t-t_ie_i-t_je_j)}{t_j} - \mathbf{D}_{x_j(\overline{x}_{M|\{i,j\}})} f(0)\Big]dt + \mathbf{D}_{x_j(\overline{x}_{M|\{i,j\}})} f(0).$$

whence we easily arrive at

$$\lim_{(x,x_{k+1})\xrightarrow[x_j]{\wedge}0} I_2 = \mathbf{D}_{x_j(\overline{x}_{M|\{i,j\}})} f(0) = \frac{\partial f(0)}{\partial x_j}.$$

Finally, we have

$$\lim_{(x,x_{k+1})\xrightarrow[x_ix_j]{\wedge}0} \frac{\partial U(f;x,x_{k+1})}{\partial x_j} = \frac{\partial f(0)}{\partial x_j}.$$

The theorem is proved. □

Analogously, we prove the following

Theorem 4.2.4. *If at the point x^0 there exist finite derivatives*

$$\mathbf{D}_{x_1(x_2,x_3,\dots,x_k)} f(x^0),\quad \mathbf{D}_{x_2(x_3,\dots,x_k)} f(x^0), \dots \mathbf{D}_{x_{k-1}(x_k)} f(x^0),\, f'_{x_k}(x^0),$$

then

$$\lim_{(x,x_{k+1})\xrightarrow[x_1]{\wedge}(x^0,0)} \frac{\partial U(f;x,x_{k+1})}{\partial x_1} = \frac{\partial f(x^0)}{\partial x_1},$$

$$\lim_{(x,x_{k+1})\xrightarrow[x_1,x_2]{\wedge}(x^0,0)} \frac{\partial U(f;x,x_{k+1})}{\partial x_2} = \frac{\partial f(x^0)}{\partial x_2},$$

. . . .

$$\lim_{(x,x_{k+1})\xrightarrow{\wedge}(x^0,0)} \frac{\partial U(f;x,x_{k+1})}{\partial x_k} = \frac{\partial f(x^0)}{\partial x_k}.$$

Corollary. *If at the point x^0 there exist finite derivatives*

$$\mathbf{D}_{x_1(\overline{x}_{M|1})} f(x^0),\quad \mathbf{D}_{x_2(\overline{x}_{M|\{1,2\}})} f(x^0) f(x^0), \dots \mathbf{D}_{x_{k-1}(x_k)} f(x^0),\, f'_{x_k}(x^0),$$

then

$$\lim_{(x,x_{k+1})\xrightarrow{\wedge}(x^0,0)} d_x U(f;x,x_{k+1}) = df(x^0).$$

Theorem 4.2.5.
(a) *If at the point* x^0 *there exists a finite derivative* $\mathbf{D}^*_{x_i(\overline{x}_{M|i})}f(x^0)$*, then*

$$\lim_{(x-x_ie_i+x_i^0e_i,x_{k+1})\to(x^0,0)}\frac{\partial U(f;x-x_ie_i+x_i^0e_i,x_{k+1})}{\partial x_i}=\mathbf{D}^*_{x_i}f(x^0).$$

(b) *There exists a continuous function* $g(x)$ *such that* $\mathbf{D}^*_{x_i(\overline{x}_{M|i})}g(x^0)=0$*, but the limit*

$$\lim_{(x,x_{k+1})\overset{\wedge}{\longrightarrow}(x^0,0)}\frac{\partial U(g;x,x_{k+1})}{\partial x_i}$$

does not exist.

Proof of Item (a). Let $x^0=0$. We have

$$\frac{\partial U(f;x,x_{k+1})}{\partial x_i}=C_kx_{k+1}\int\limits_{R^k}\frac{(t_i-x_i)f(t)dt}{\left(|t-x|^2+x_{k+1}^2\right)^{\frac{k+3}{2}}},$$

hence

$$\frac{\partial U(f;x-x_ie_i,x_{k+1})}{\partial x_i}=C_kx_{k+1}\int\limits_{R^k}\frac{t_if(t)dt}{\left(|t-x+x_ie_i|^2+x_{k+1}^2\right)^{\frac{k+3}{2}}}.$$

Transforming $t_1-x_1=\tau_1$, $t_2-x_2=\tau_2,\dots,t_i=\tau_i,\dots,t_k-x_k=\tau_k$ the latter equality yields

$$\begin{aligned}\frac{\partial U(f;x-x_ie_i,x_{k+1})}{\partial x_i}&=C_kx_{k+1}\int\limits_{R^k}\frac{t_1f(t_1+x_1,\dots,t_i,\dots,t_k+x_k)dt}{\left(|t|^2+x_{k+1}^2\right)^{\frac{k+3}{2}}}\\&=C_kx_{k+1}\int\limits_{R^k}\frac{t_if(t+x-x_ie_i)dt}{\left(|t|^2+x_{k+1}^2\right)^{\frac{k+3}{2}}}.\end{aligned}\tag{2.4}$$

By means of the substitution $t_i=-\tau_i$, from (2.4) we get

$$\begin{aligned}\frac{\partial U(f;x-x_ie_i,x_{k+1})}{\partial x_i}&=-C_kx_{k+1}\int\limits_{R^k}\frac{t_if(t_i+x_1,\dots,-t_i,\dots,t_k+x_k)dt}{\left(|t|^2+x_{k+1}^2\right)^{\frac{k+3}{2}}}\\&=-C_kx_{k+1}\int\limits_{R^k}\frac{t_if(t+x-x_ie_i-2t_ie_i)dt}{\left(|t|^2+x_{k+1}^2\right)^{\frac{k+3}{2}}}.\end{aligned}\tag{2.5}$$

(2.4) and (2.5) result in

$$\frac{\partial U(f;x-x_ie_i,x_{k+1})}{\partial x_i}$$

$$= C_k x_{k+1} \int\limits_{R^k} \frac{t_i}{\left(|t|^2 + x_{k+1}^2\right)^{\frac{k+3}{2}}} \cdot \frac{f(t + x - x_i e_i) - f(t + x - x_i e_i - 2t_i e_i)}{2} dt.$$

By virtue of Lemma 4.2.1, this equality leads to

$$\begin{aligned} &\frac{\partial U(f; x - x_i e_i, x_{k+1})}{\partial x_i} - \mathbf{D}^*_{x_i(\overline{x}_{M|i})} f(0) \\ &\qquad = C_k x_{k+1} \int\limits_{R^k} \frac{t_i}{\left(|t|^2 + x_{k+1}^2\right)^{\frac{k+3}{2}}} \\ &\times \left[\frac{f(t + x - x_i e_i) - f(t + x - x_i e_i - 2t_i e_i)}{2t_i} - \mathbf{D}^*_{x_i(\overline{x}_{M|i})} f(0) \right] dt. \end{aligned} \tag{2.6}$$

Reasoning as in proving Theorem 4.2.1, from (2.6) we find that Item (a) of Theorem 4.2.5 is valid.

Proof of Item (b). Assume $\mathbf{D}_1 = [0, 1; 0, 1]$, $\mathbf{D}_2 = [-1, 0; 0, 1]$. Let

$$g(t_1, t_2) = \begin{cases} \sqrt{t_1 \sqrt{t_2}}, & \text{when } (t_1, t_2) \in \mathbf{D}_1, \\ \sqrt{-t_1 \sqrt{t_2}}, & \text{when } (t_1, t_2) \in \mathbf{D}_2, \\ 0, & \text{when } t_2 \leq 0, \end{cases}$$

and on the set $R_+^2 | (\mathbf{D}_1 \cup \mathbf{D}_2)$ we extend $g(t_1, t_2)$ continuously so that $g \in L(R^2)$. It is easy to verify that $\mathbf{D}^*_{t_1(t_2)} g(0) = 0$. Let $x_1^0 = x_2^0 = 0$ and $(x_1, x_2, x_3) \to (0, 0, 0)$ so that $x_2 = 0$ and $x_3 = x_1$. Then for the above-constructed function,

$$\frac{\partial U(g; x_1, x_2, x_3)}{\partial x_1} = \frac{3x_3}{2\pi} \int\limits_{R^2} \frac{(t_1 - x_1) g(t_1, t_2) dt_1 dt_2}{[(t_1 - x_1)^2 + (t_2 - x_2)^2 + x_3^2]^{\frac{5}{2}}}$$

$$= C x_3 \left\{ \int\limits_{-1}^{0} \int\limits_{0}^{1} \frac{(t_1 - x_1) \sqrt{-t_1 \sqrt{t_2}} dt_1 dt_2}{[(t_1 - x_1)^2 + t_2^2 + x_3^2]^{\frac{5}{2}}} + \int\limits_{0}^{1} \int\limits_{0}^{1} \frac{(t_1 - x_1) \sqrt{t_1 \sqrt{t_2}} dt_1 dt_2}{[(t_1 - x_1)^2 + t_2^2 + x_3^2]^{\frac{5}{2}}} \right\} + o(1)$$

$$= C x_1 \left[- \int\limits_{x_1}^{1 + x_1} \int\limits_{0}^{1} \frac{t_1 \sqrt{(t_1 - x_1) \sqrt{t_2}}}{(t_1^2 + t_2^2 + x_1^2)^{\frac{5}{2}}} dt_1 dt_2 + \int\limits_{-x_1}^{1 - x_1} \int\limits_{0}^{1} \frac{t_1 \sqrt{(t_1 + x_1) \sqrt{t_2}}}{(t_1^2 + t_2^2 + x_1^2)^{\frac{5}{2}}} dt_1 dt_2 \right] + o(1)$$

$$\begin{aligned} &= C x_1 \left\{ \int\limits_{-x_1}^{x_1} \int\limits_{0}^{1} \frac{t_1 \sqrt{(t_1 + x_1) \sqrt{t_2}}}{(t_1^2 + t_2^2 + x_1^2)^{\frac{5}{2}}} dt_1 dt_2 \right. \\ &\left. + \int\limits_{x_1}^{1 - x_1} \int\limits_{0}^{1} \frac{t_1 \left[\sqrt{(t_1 + x_1) \sqrt{t_2}} - \sqrt{(t_1 - x_1) \sqrt{t_2}} \right]}{(t_1^2 + t_2^2 + x_1^2)^{\frac{5}{2}}} dt_1 dt_2 \right\} + o(1) \\ &= C x_1 (I_1 + I_2) + o(1), \end{aligned}$$

where

$$I_1=\int\limits_0^{x_1}\int\limits_0^1\frac{t_1\big[\sqrt{(t_1+x_1)\sqrt{t_2}}-\sqrt{(x_1-t_1)\sqrt{t_2}}\big]}{(t_1^2+t_2^2+x_1^2)^{\frac52}}dt_1dt_2>0,$$

$$I_2=\int\limits_{x_1}^{1-x_1}\int\limits_0^1\frac{t_1\sqrt[4]{t_2}\big(\sqrt{t_1+x_1}-\sqrt{t_1-x_1}\big)}{(t_1^2+t_2^2+x_1^2)^{\frac52}}dt_1dt_2$$

$$>\int\limits_{x_1}^{2x_1}\int\limits_{x_1}^{2x_1}\frac{t_1\sqrt[4]{t_2}\big(\sqrt{t_1+x_1}-\sqrt{t_1-x_1}\big)}{(t_1^2+t_2^2+x_1^2)^{\frac52}}dt_1dt_2,$$

$$>\int\limits_{x_1}^{2x_1}\int\limits_{x_1}^{2x_1}\frac{x_1\sqrt[4]{x_1}(\sqrt{2x_1}-\sqrt{x_1})}{(9x_1^2)^{\frac52}}dt_1dt_2=\frac{\sqrt2-1}{128}\cdot\frac{1}{\sqrt[4]{x_1^5}}.$$

Thus along the chosen path we have

$$\frac{\partial U(g;x_1,0,x_1)}{\partial x_1}>\frac{C}{\sqrt[4]{x_1}},$$

whence $\dfrac{\partial U(g;x_1,0,x_1)}{\partial x_1}\to+\infty$ as $(x_1,x_2,x_3)\to(0,0,0)$ along the chosen path.
The theorem is proved. □

Analogously, we prove the following

Theorem 4.2.6.
(a) *If at the point x^0 there exists the finite derivative $\overline{\mathbf{D}}^*_{x_i(x)}f(x^0)$, $i=\overline{1,k}$, then*

$$\lim_{(x,x_{k+1})\to(x^0,0)}\frac{\partial U(f;x,x_{k+1})}{\partial x_i}=\mathbf{D}^*_{x_i}f(x^0).$$

(b) *There exists a continuous function $g(x)$ such that for any $B\subset M$, $m(B)<k$ we have $\mathbf{D}_{x_i(\overline{x}_B)}g(0)=0$, $i=\overline{1,k}$, but the limits*

$$\lim_{x_{k+1}\to0+}\frac{\partial U(g;0,x_{k+1})}{\partial x_i}$$

do not exist.

Item (a) of Theorem 4.2.1 is a corollary of Item (a) of Theorem 4.2.6.

Theorem 4.2.7. (a) *If at the point x^0 the function f has a total differential $df(x^0)$, then*

$$\lim_{(x,,x_{k+1})\overset{\wedge}{\longrightarrow}(x^0,0)}d_xU(f;x,x_{k+1})=df(x^0).\tag{2.7}$$

(b) *There exists a continuous function g whose all partial derivatives at the point x^0 are of any order, however, but the limits*

$$\lim_{x_{k+1}\to 0+} \frac{\partial U(g; x^0, x_{k+1})}{\partial x_i}, \quad i = \overline{1, k},$$

do not exist.

Proof of Item (a). By Statements 1) and 2), from Lemma 4.2.1 we have (here $x^0 = 0$)

$$\frac{\partial U(f; x, x_{k+1})}{\partial x_i} - \frac{\partial f(0)}{\partial x_i} = C_k x_{k+1} \int\limits_{R^k} \frac{(t_i - x_i)\sum_{v=1}^k |t_v|}{(|t-x|^2 + x_{k+1}^2)^{\frac{k+3}{2}}}$$
$$\times \frac{f(t) - f(0) - \sum_{v=1}^k \frac{\partial f(0)}{\partial x_v} t_v}{\sum_{v=1}^k |t_v|} dt.$$

In view of Statements 3), 4) and 5) of Lemma 4.2.1, the last equality yields

$$\lim_{(x, x_{k+1}) \overset{\wedge}{\longrightarrow} 0} \frac{\partial U(f; x, x_{k+1})}{\partial x_i} = \frac{\partial f(0)}{\partial x_i}, \quad i = \overline{1, k}.$$

Thus the equality (2.7) is fulfilled.

Proof of Item (b). Consider the function

$$g(t_1, t_2) = \begin{cases} \sqrt[4]{(2t_1 - t_2)\left(t_2 - \frac{1}{2}t_1\right)}, & \text{when } (t_1, t_2) \in D \\ & = \left\{(t_1, t_2) : 0 \le t_1 < \infty; \frac{1}{2}t_1 \le t_2 \le 2t_1\right\}, \\ 0, & \text{when } (t_1, t_2) \in CD. \end{cases}$$

This function is continuous in R^2, its all partial derivatives at the point $(0, 0)$ are of any order and all of them are equal to zero, but

$$\frac{\partial U(g; 0, 0, x_3)}{\partial x_1} = \frac{3x_3}{2\pi} \int\limits_0^\infty dt_1 \int\limits_{\frac{1}{2}t_1}^{2t_1} \frac{t_1 \sqrt[4]{(2t_1 - t_2)(t_2 - \frac{1}{2}t_1)}}{(t_1^2 + t_2^2 + x_3^2)^{\frac{5}{2}}} dt_2$$
$$> Cx_3 \int\limits_{x_3}^{2x_3} t_1 dt_1 \int\limits_{t_1}^{\frac{3}{2}t_1} \frac{\sqrt[4]{(2t_1 - t_2)(t_2 - \frac{1}{2}t_1)}}{(t_1^2 + t_2^2 + x_3^2)^{\frac{5}{2}}} dt_2$$
$$> Cx_3 \int\limits_{x_3}^{2x_3} t_1 dt_1 \int\limits_{t_1}^{\frac{3}{2}t_1} \frac{\sqrt[4]{(2t_1 - \frac{3}{2}t_1)(t_1 - \frac{1}{2}t_1)}}{\left(\frac{13}{4}t_1^2 + x_3^2\right)^{\frac{5}{2}}} dt_2$$

$$> Cx_3 \int_{x_3}^{2x_3} t_1 dt_1 \int_{t_1}^{\frac{3}{2}t_1} \frac{\sqrt[4]{t_1^2}dt}{x_3^{\frac{5}{3}}} = \frac{C}{x_3^{\frac{4}{3}}} \int_{x_3}^{2x_3} t_1^{\frac{5}{2}} dt_1 = \frac{C}{\sqrt{x_3}} \to +\infty, \quad \text{as} \quad x_3 \to 0+.$$

The theorem is proved. □

4.3 Generalized Partial Second Order Derivatives for a Function of Several Variables

Let $u \in R$, $v \in R$. Consider the following derivatives of the function $f(x) = f(x_1, x_2, \ldots, x_k)$:

1. Let $\delta > 0$, $V_\delta = \prod\limits_{v=1}^{k} [x_v^0 - \delta; x_v^0 + \delta]$ and $f'_{x_i}(x) \in L(V_\delta)$.

Denote the limit

$$\lim_{(u,\overline{x}_B)\to(0,\overline{x}^0_B)} \frac{f(x_B + x^0_{B'} + ue_i) - f(x_B + x^0_{B'}) - f'_{x_i}(x_B + x^0_{B'})u}{\frac{1}{2}u^2}$$

a) by $\overline{f}''_{x_i}(x^0)$ if $B = \varnothing$;
b) by $\overline{f}''_{x_i(\overline{x}_B)}(x^0)$ if $i \in B'$;
c) by $\widetilde{f}''_{x_i(\overline{x}_B)}(x^0)$ if $i \in B$.

2. Denote the limit

$$\lim_{(u,\overline{x}_B)\to(0,\overline{x}^0_B)} \frac{f(x_B + x^0_{B'} + ue_i) + f(x_B + x^0_{B'} - ue_i) - 2f(x_B + x^0_{B'})}{u^2}$$

a) by $\mathbf{D}^2_{x_i} f(x^0)$, if $B = \varnothing$;
b) by $\mathbf{D}^2_{x_i(\overline{x}_B)} f(x^0)$, if $i \in B'$;
c) by $\overline{\mathbf{D}}^2_{x_i(\overline{x}_B)} f(x^0)$, if $i \in B$.

3. Denote the limit

$$\lim_{\substack{(u,v)\to(0,0)\\ \overline{x}_B\to\overline{x}^0_B}} \frac{f(x_B + x^0_{B'} + ue_i + ve_j) - f(x_B + x^0_{B'} + ue_i) - f(x_B + x^0_{B'} + ve_j) + f(x_B + x^0_{B'})}{uv}$$

a) by $\mathbf{D}^2_{x_i x_j} f(x^0)$, if $B = \varnothing$;
b) by $\mathbf{D}^2_{x_i x_j(\overline{x}_B)} f(x^0)$, if $\{i, j\} \subset B'$;
c) by $\mathbf{D}^2_{[x_i x_j](\overline{x}_B)} f(x^0)$, if $\{i, j\} \subset B$;
d) by $\mathbf{D}^2_{[x_i]x_j(\overline{x}_B)} f(x^0)$, if $i \in B$, $j \in B'$;

4. Denote the limit

$$\lim_{\substack{(u,v)\to(0,0)\\ \overline{x}_B\to\overline{x}^0_B}}\Big[\frac{f(x_B+x^0_{B'}+ue_i+ve_j)-f(x_B+x^0_{B'}+ue_i-ve_j)}{4uv}$$

$$-\frac{f(x_B+x^0_{B'}-ue_i+ve_j)-f(x_B+x^0_{B'}-ue_i-ve_j)}{4uv}\Big]$$

a) by $\mathbf{D}^*_{x_ix_j}f(x^0)$ if $B=\varnothing$;
b) by $\mathbf{D}^*_{x_ix_j(\overline{x}_B)}f(x^0)$ if $\{i,j\}\subset B'$;
c) by $\mathbf{D}^*_{[x_ix_j](\overline{x}_B)}f(x^0)$ if $\{i,j\}\subset B$;
d) $\mathbf{D}^*_{[x_i]x_j(\overline{x}_B)}f(x^0)$ if $i\in B$, $j\in B'$;
The following statements are valid.

1) For any $B\subset M$, the existence of $\widetilde{f}''_{x_i(\overline{x}_B)}(x^0)$ implies the existence of $\overline{\mathbf{D}}_{x_i(\overline{x}_B)}(x^0)$ and

$$\widetilde{f}''_{x_i(\overline{x}_B)}(x^0)=\overline{\mathbf{D}}^2_{x_i(\overline{x}_B)}(x^0)=\widetilde{f}''_{x_i}(x^0)=\overline{\mathbf{D}}^2_{x_i}(x^0).$$

This follows from the equality

$$\frac{f(x_B+x^0_{B'}+ue_i)+f(x_B+x^0_{B'}-ue_i)-2f(x_B+x^0_{B'})}{u^2}$$

$$=\frac{1}{2}\Big[\frac{f(x_B+x^0_{B'}+ue_i)+f(x_B+x^0_{B'})-f'_{x_i}(x_B+x^0_{B'})u}{\frac{1}{2}u^2}$$

$$+\frac{f(x_B+x^0_{B'}-ue_i)-f(x_B+x^0_{B'})-f'_{x_i}(x_B+x^0_{B'})(-u)}{\frac{1}{2}u^2}\Big].$$

2) If there exists $f''_{x_i}(x^0)$, then there exist $\overline{f}''_{x_i}(x^0)$ and $D^2_{x_i}f(x^0)$ and they have one and the same value.

3) If there exists a partial derivative $f''_{x_i}(x)$ in the neighborhood of the point x^0 and it is continuous at x^0, then there also exists $\widetilde{f}''_{x_i(x)}(x^0)$ (therefore there exists $\overline{\mathbf{D}}^2_{x_i(x)}f(x^0)$ too) and

$$\widetilde{f}''_{x_i(x)}(x^0)=\overline{\mathbf{D}}^2_{x_i(x)}f(x^0)=f''_{x_i}(x^0).$$

Indeed, if to the functions $f(x+ue_i)-f(x)-f'_{x_i}(x)u$ and $\frac{1}{2}u^2$ we apply the Cauchy formula with respect to u, then we obtain

$$\frac{f(x+ue_i)-f(x)-f'_{x_i}(x)u}{\frac{1}{2}u^2}=\frac{f'(x_0+\theta(x)ue_i)-f'_{x_i}(x)}{\theta(x)u},\quad 0<\theta<1.$$

Using now the Lagrange formula, we find that

$$\frac{f(x+ue_i)-f(x)-f'_{x_i}(x)u}{\frac{1}{2}u^2}=\frac{\theta(x)uf''(x+\theta_1\theta ue_i)}{\theta(x)u}$$
$$=f''(x+\theta_1\theta ue_i),\quad 0<\theta_1<1.$$

whence taking into account that $f''_{x_i}(x)$ is continuous, we obtain that Statement 3) is valid.

It should be noted that the continuity of the partial derivative $f''_{x_i^2}(x)$ at the point x^0 is only the sufficient condition for the existence of derivatives $f''_{x_i(x_B)}(x^0)$ and $\overline{\mathbf{D}}^2_{x_i(x_B)}f(x^0)$ for any $B\subset M$.

4) For any $B\subset M$, the existence of $\overline{\mathbf{D}}^2_{x_i(\overline{x}_B)}f(x^0)$ implies the existence of $\overline{\mathbf{D}}^2_{x_i(x_{B|i})}f(x^0)$ and they have one and the same value.

5) If there exists a derivative $f''_{x_ix_j}(x)$ in the neighborhood of the point x^0 and it is continuous at x^0, then there exist $\overline{\mathbf{D}}^2_{[x_ix_j](x)}f(x^0)$ and $\overline{\mathbf{D}}^2_{[x_ix_j](x)}f(x^0)=f''_{x_i,x_j}(x^0)$.

The continuity of $f''_{x_i,x_j}(x)$ at the point x_0 is the sufficient condition for the existence of $\overline{\mathbf{D}}^2_{[x_ix_j](x)}f(x^0)$.

6) If for the function $f(x)$ at the point x^0 there exists $\mathbf{D}^2_{x_ix_j}f(x^0)$, then at this point there also exists $\mathbf{D}^*_{x_ix_j}f(x^0)$ and they have one and the same value.

4.4 The Boundary Properties of Partial Second Order Derivatives of the Poisson Integral for a Half-Space R^{k+1}_+ $(k>1)$

It can be easily verified that

$$\frac{\partial^2U(f;x,x_{k+1})}{\partial x_i^2}=\frac{\Gamma\left(\frac{k+1}{2}\right)}{\pi^{\frac{k+1}{2}}}\int\limits_{R^k}\frac{\partial^2P(t-x,x_{k+1})}{\partial x_i^2}f(t)dt$$

$$=\frac{(k+1)x_{k+1}\Gamma\left(\frac{k+1}{2}\right)}{\pi^{\frac{k+1}{2}}}\int\limits_{R^k}\frac{(k+3)(t_i-x_i)^2-|t-x|^2-x_{k+1}^2}{(|t-x|^2+x_{k+1}^2)^{\frac{k+5}{2}}}f(t)dt.$$

Lemma 4.4.1. *For every $(x,x_{k+1})\in R^{k+1}_+$, the following statements are valid:*

1) $\int\limits_{-\infty}^{\infty}\frac{\partial^2P(t-x,x_{k+1})}{\partial x_i^2}t_i^v\,dt_i=0$, $v=0,1$;

2) $\int\limits_{R^k}\frac{\partial^2P(t-x,x_{k+1})}{\partial x_i^2}t_i^v f(t-t_ie_i)dt=0$, $v=0,1$;

3) $I = \dfrac{\Gamma\left(\dfrac{k+1}{2}\right)}{\pi^{\frac{k+1}{2}}} \int\limits_{R^k} \frac{\partial^2 P(t-x,x_{k+1})}{\partial x_i^2} \cdot \dfrac{t_i^2}{2!} dt = 1;$

4) $\left| \int\limits_{R^k} \dfrac{\partial^2 P(t, x_{k+1})}{\partial t_i^2} \right| |t^2| dt = O(1);$

5) $\left| \int\limits_{R^k} \dfrac{\partial^2 P(t-x, x_{k+1})}{\partial t_i^2} \right| t_i^2 dt = O(1),\ for\ \dfrac{x_{k+1}}{|x_i|} \geq C > 0;$

6) $\lim\limits_{x_{k+1}\to 0} \sup\limits_{|t|\geq\delta>0} \left| \frac{\partial^2 P(t,x)}{\partial t_i^2} \right| |t|^2 dt = 0.$

Proof. For $v = 0$ the validity of Statement 1) follows from the fact that the integral

$$\int\limits_{R^k} P(t-x, x_{k+1}) dt = \int\limits_{R_k} P(t, x_{k+1}) dt = \frac{\pi^{\frac{k+1}{2}}}{\Gamma\left(\frac{k+1}{2}\right)}$$

does not depend on x and therefore

$$\int\limits_{-\infty}^{\infty} \frac{\partial^2 P(t-x, x_{k+1})}{\partial x_i^2} dt_i = \frac{\partial^2}{\partial x_i^2} \int\limits_{-\infty}^{\infty} P(t-x, x_{k+1}) dt_i = 0.$$

For $v = 1$, the validity of Statement 1) follows from the fact that the subintegrand is odd with respect to t_i.

Statement 2) follows from Statement 1).

Consider now the integral

$$I = \frac{\Gamma\left(\frac{k+1}{2}\right)}{\pi^{\frac{k+1}{2}}} \int\limits_{R^k} \frac{\partial^2 P(t-x, x_{k+1})}{\partial x_i^2} \cdot \frac{t_i^2}{2!} dt$$

$$= \frac{(k+1)\Gamma\left(\frac{k+1}{2}\right) x_{k+1}}{\pi^{\frac{k+1}{2}}} \int\limits_{R^k} \frac{(k+3)(t_i - x_i)^2 - |t-x|^2 - x_{k+1}^2}{\left(|t-x|^2 + x_{k+1}^2\right)^{\frac{k+5}{2}}} t_i^2 dt.$$

Passing to the spherical coordinates $(\rho, \theta_1, \theta_2, \ldots, \theta_{k-2}, \varphi)$, using the equalities ([10], p.311, No 10 and p. 1023, No 8.756)

$$\int\limits_0^{\pi} \sin^{\mu-1} \theta d\theta = \frac{\Gamma\left(\frac{\mu}{2}\right)}{\Gamma\left(\frac{\mu+1}{2}\right)} \sqrt{\pi},$$

$$\int\limits_0^{\infty} \frac{\rho^{k+1} d\rho}{(1+\rho^2)^{\frac{k+5}{2}}} = \frac{k\Gamma\left(\frac{k}{2}\right)\sqrt{\pi}}{2(k+1)(k+3)\Gamma\left(\frac{k+1}{2}\right)},$$

$$\int\limits_0^\infty \frac{\rho^{k+3}d\rho}{(1+\rho^2)^{\frac{k+5}{2}}} = \frac{k(k+2)\Gamma\left(\frac{k}{2}\right)\sqrt{\pi}}{2(k+1)(k+3)\Gamma\left(\frac{k+1}{2}\right)},$$

and performing calculations analogous to those we made in proving Lemma 4.2.1, we obtain

$$I = 1.$$

Let us now proceed prove the validity of Statement 4). We have

$$\int\limits_{R^k} \left|\frac{\partial^2 P(t,x)}{\partial t_i^2}\right| |t|^2 dt = Cx_{_{k+1}} \int\limits_{R^k} \frac{|(k+3)t_i^2 - |t|^2 - x_{k+1}^2|}{(|t|^2 + x_{k+1}^2)^{\frac{k+5}{2}}} |t|^2 dt$$

$$\leq Cx_{_{k+1}} \int\limits_{R^k} \frac{I(t_1,t_2,\dots,t_k,x_{k+1})}{(|t|^2 + x_{k+1}^2)^{\frac{k+5}{2}}} |t|^2 dt,$$

where $I(t_1,t_2,\dots,t_k,x_{k+1})$ is a homogeneous polynomial of degree 2 with respect to $(t_1,t_2,\dots,t_k,x_{k+1})$.

Passing to the spherical coordinates, we have

$$\int\limits_{R^k} \left|\frac{\partial^2 P(t,x_{k+1})}{\partial t_i^2}\right| |t|^2 dt \leq Cx_{k+1} \int\limits_0^\infty \frac{T(\rho,x_{k+1})\rho^{k+1}}{(\rho^2 + x_{k+1}^2)^{\frac{k+5}{2}}} d\rho,$$

where $T(\rho,x_{k+1})$ is a homogeneous polynomial of degree 2 in (ρ,x_{k+1}). Using the substitution $\rho = x_{k+1}\rho_1$, we obtain

$$\int\limits_{R^k} \left|\frac{\partial^2 P(t,x_{k+1})}{\partial t_i^2}\right| |t|^2 dt \leq C \sum_{v=0}^{2} \int\limits_0^\infty \frac{\rho_1^{k+1+v} d\rho_1}{(1+\rho_1^2)^{\frac{k+5}{2}}} = O(1).$$

The validity of Statement 5) follows from the inequality

$$\int\limits_{R^k} \left|\frac{\partial^2 P(t-x,x_{k+1})}{\partial t_i^2}\right| t_i^2 dt \leq C_k x_{k+1} \int\limits_{R^k} \frac{(k+3)(t_i-x_i)^2 + |t-x|^2 + x_{k+1}^2}{(|t-x|^2 + x_{k+1}^2)^{\frac{k+5}{2}}} t_i^2 dt$$

$$= C_k x_{k+1} \int\limits_{R^k} \frac{(k+3)t_i^2 + |t|^2 + x_{k+1}^2}{(|t|^2 + x_{k+1}^2)^{\frac{k+5}{2}}} (t_i + x_i)^2 dt,$$

if we continue the reasoning as in proving Statement 4) and take into account the condition $\dfrac{x_{k+1}}{|x_i|} \geq C > 0$. Statement 6) is obvious. □

Theorem 4.4.1. (a) *If at the point x^0 there exists a finite derivative $\overline{\mathbf{D}}^2_{x_i(x)}f(x^0)$, then*

$$\lim_{(x,x_{k+1})\to(x^0,0)} \frac{\partial^2 U(f;x,x_{k+1})}{\partial x_i^2} = \mathbf{D}^2_{x_i}f(x^0) = \frac{\partial^2 f(x^0)}{\partial x_i^2}, \tag{4.1}$$

(b) *There exists a continuous function $f \in L(R^k)$ such that for every $B \subset M$, $m(B) < k$ all derivatives $\mathbf{D}^2_{x_i(\overline{x}_B)}f(0) = 0$, $i = \overline{1,k}$, but the limits*

$$\lim_{x_{k+1}\to 0+} \frac{\partial^2 U(f;0,x_{k+1})}{\partial x_i^2}$$

do not exist.

Proof of Item (a). Let $x^0 = 0$, $C_k = \frac{(k+1)\Gamma\left(\frac{k+1}{2}\right)}{\pi^{\frac{k+1}{2}}}$.

We have

$$\frac{\partial^2 U(f;x,x_{k+1})}{\partial x_i^2} = C_k \int\limits_{R^k} \frac{\partial^2 P(t-x,x_{k+1})}{\partial x_i^2} f(t)dt$$
$$= C_k \int\limits_{R^k} \frac{\partial^2 P(t,x_{k+1})}{\partial x_i^2} f(x+t)dt. \tag{4.2}$$

From the integral (4.2), using the substitution $t_i = -\tau_i$, we obtain

$$\frac{\partial^2 U(f;x,x_{k+1})}{\partial x_i^2} = C_k \int\limits_{R^k} \frac{\partial^2 P(t,x_{k+1})}{\partial x_i^2} f(x+t-2t_ie_i)dt. \tag{4.3}$$

(4.2) and (4.3) result in

$$\frac{\partial^2 U(f;x,x_{k+1})}{\partial x_i^2} = \frac{1}{2}C_k \int\limits_{R^k} \frac{\partial^2 P(t,x_{k+1})}{\partial x_i^2} [f(x+t)+f(x+t-2t_ie_i)]dt,$$

which by virtue of Statements 2) and 3) of Lemma 4.4.1, results in

$$\frac{\partial^2 U(f;x,x_{k+1})}{\partial x_i^2} - \overline{\mathbf{D}}^2_{x_i(x)}f(0) = \frac{1}{2}C_k \int\limits_{R^k} \frac{\partial^2 P(t,x_{k+1})}{\partial x_i^2} t_i^2$$
$$\times \left[\frac{f(x+t)+f(x+t-2t_ie_i)-2f(x+t-t_ie_i)}{t_i^2} - \overline{\mathbf{D}}^2_{x_i(x)}f(0)\right]dt. \tag{4.4}$$

Taking into account Statements 4) and 6) of Lemma 4.4.1, it follows from (4.4) that the equality (4.1) is valid.

Proof of Item (b). Assume $\mathbf{D}=[0\le t_1<\infty;0\le t_2<\infty;0\le t_3<\infty]$. Let

$$f(t)\begin{cases}\sqrt[3]{t_1t_2t_3}, & \text{if } (t_1,t_2,t_3)\in D,\\ 0, & \text{if } (t_1,t_2,t_3)\in CD.\end{cases}$$

It can be easily verified that

$$\overline{\mathbf{D}}^2_{x_i(x_i,x_j)}f(0)=0,\quad i,j=1,2,3;\quad i\neq j.$$

However, for the considered function,

$$\frac{\partial^2U(f;0,x_4)}{\partial x_1^2}=\frac{4x_4}{\pi^2}\int\limits_{\mathbf{D}}\frac{6t_1^2-|t|^2-x_4^2}{(|t|^2+x_4^2)^4}\sqrt[3]{t_1t_2t_3}dt$$

$$=\frac{4x_4}{\pi^2}\int\limits_0^\infty\int\limits_0^{\frac{\pi}{2}}\int\limits_0^{\frac{\pi}{2}}\frac{6\rho^2\sin^2\theta\cos^2\varphi-\rho^2-x_4^2}{(\rho^2+x_4^2)^4}$$

$$\times\sqrt[3]{\rho^3\sin^2\theta\cos\theta\sin\varphi\cos\varphi}\rho^2\sin\theta d\rho d\theta d\varphi$$

$$=\frac{C}{x_4}\Big(4\int\limits_0^{\frac{\pi}{2}}\sqrt[3]{\sin^2\theta\cos\theta}\sin^3\theta d\theta\int\limits_0^{\frac{\pi}{2}}\sqrt[3]{\sin2\varphi}\sin^2\varphi d\varphi$$

$$-\int\limits_0^{\frac{\pi}{2}}\sqrt[3]{\sin^2\theta\cos\theta}\sin\varphi d\theta\int\limits_0^{\frac{\pi}{2}}\sqrt[3]{sin2\varphi d\varphi}\Big).$$

One can verify that

$$\left.\begin{aligned}\int\limits_0^{\frac{\pi}{2}}\sqrt[3]{\sin^2\theta\cos\theta}\sin\theta d\theta=\frac{1}{6}\Gamma\Big(\frac{1}{3}\Big)\Gamma\Big(\frac{2}{3}\Big)\\ \int\limits_0^{\frac{\pi}{2}}\sqrt[3]{\sin^2\theta\cos\theta}\sin^3\theta d\theta=\frac{1}{9}\Gamma\Big(\frac{1}{3}\Big)\Gamma\Big(\frac{2}{3}\Big)\end{aligned}\right\}.\tag{4.5}$$

Taking into account the equalities (4.5), we have

$$\frac{\partial^2U(f;0,x_4)}{\partial x_1^2}=\frac{C}{x_4}\int\limits_0^{\frac{\pi}{2}}\Big(\sin^2\varphi-\frac{3}{8}\Big)\sqrt[3]{\sin2\varphi}d\varphi$$

$$=\frac{C}{x_4}\Bigg(\int\limits_0^{\arcsin\sqrt{\frac{3}{8}}}+\int\limits_{\arcsin\sqrt{\frac{3}{8}}}^{\frac{\pi}{2}-\arcsin\sqrt{\frac{3}{8}}}+\int\limits_{\frac{\pi}{2}-\arcsin\sqrt{\frac{3}{8}}}^{\frac{\pi}{2}}\Bigg)=\frac{C}{x_4}(I_1+I_2+I_3).$$

Clearly, $I_1<0$, $I_2>0$, $I_3>0$. Furthermore,

$$I_3=\int\limits_0^{\arcsin\sqrt{\frac{3}{8}}}\left(\cos^2\varphi-\frac{3}{8}\right)\sqrt[3]{\sin 2\varphi}d\varphi.$$

It is easily seen that $I_1+I_3>0$. Hence $I_1+I_2+I_3>0$. Finally, we have

$$\frac{\partial^2U(f;0,x_4)}{\partial x_1^2}\to+\infty \quad\text{as}\quad x_4\to 0+.$$

Theorem 4.4.1 is proved. □

Corollary 4.4.1. *If at the point x^0 there exists a finite $\widetilde{f}''_{x_i(x)}(x^0)$, then the equality* (4.1) *is fulfilled.*

Corollary 4.4.2. *If the function f has a continuous partial derivative $\frac{\partial^2 f(x)}{\partial x_i^2}$ at the point x^0, then the equality* (4.1) *is fulfilled.*

Theorem 4.4.2.

(a) *If at the point x^0 there exists a finite derivative $\mathbf{D}^2_{x_i(x_{M|i})}f(x^0)$, then*

$$\lim_{(x-x_ie_i+x_i^0e_i,x_{k+1})\to(x^0,0)}\frac{\partial^2U(f;x-x_ie_i+x_i^0e_i,x_{k+1})}{\partial x_i^2}=\mathbf{D}^2_{x_i}f(x^0).$$

(b) *There exists a function $f\in L(R^k)$ such that $\mathbf{D}^2_{x_i(x_{M|i})}f(x^0)=0$, but the limit*

$$\lim_{(x,x_{k+1})\overset{\wedge}{\longrightarrow}(x^0,0)}\frac{\partial^2U(f;x,x_{k+1})}{\partial x_i^2}$$

does not exist.

Proof of Item (a). Let $x^0=0$. We have

$$\frac{\partial^2U(f;x,x_{k+1})}{\partial x_i^2}=C_kx_{k+1}\int\limits_{R^k}\frac{(k+3)(t_i-x_i)^2-|t-x|^2-x_{k+1}^2}{(|t-x|^2+x_{k+1}^2)^{\frac{k+5}{2}}}f(t)dt,$$

whence

$$\frac{\partial^2U(f;x-x_ie_i,x_{k+1})}{\partial x_i^2}=C_kx_{k+1}\int\limits_{R^k}\frac{(k+3)t_i^2-|t-x+x_ie_i|^2-x_{k+1}^2}{(|t-x+x_ie_i|^2+x_{k+1}^2)^{\frac{k+5}{2}}}f(t)dt.$$

Using the substitution $t_1-x_1=\tau_1$, $t_2-x_2=\tau_2,\dots,t_i=\tau_i,\dots,t_k-x_k=\tau_k$ the last equality yields

$$\frac{\partial^2U(f;x-x_ie_i,x_{k+1})}{\partial x_i^2}=C_kx_{k+1}\int\limits_{R^k}\frac{(k+3)t_i^2-|t|^2-x_{k+1}^2}{(|t|^2+x_{k+1}^2)^{\frac{k+5}{2}}}$$

$$\times f(t+x-x_ie_i)dt. \tag{4.6}$$

Substituting $t_i=-\tau_i$, from (4.6) we obtain

$$\frac{\partial^2 U(f;x-x_ie_i,x_{k+1})}{\partial x_i^2}=C_kx_{k+1}\int\limits_{R^k}\frac{(k+3)t_i^2-|t|^2-x_{k+1}^2}{(|t|^2+x_{k+1}^2)^{\frac{k+5}{2}}}$$

$$\times f(t_1+x_1,t_2+x_2,\dots,-t_i,\dots,t_k+x_k)dt$$

$$=C_kx_{k+1}\int\limits_{R^k}\frac{(k+3)t_i^2-|t|^2-x_{k+1}^2}{(|t|^2+x_{k+1}^2)^{\frac{k+5}{2}}}f(t+x-2t_ie_i-x_ie_i)dt. \tag{4.7}$$

(4.6) and (4.7) result in

$$\frac{\partial^2 U(f;x-x_ie_i,x_{k+1})}{\partial x_i^2}=C_kx_{k+1}\int\limits_{R^k}\frac{(k+3)t_i^2-|t|^2-x_{k+1}^2}{(|t|^2+x_{k+1}^2)^{\frac{k+5}{2}}}$$

$$\times\frac{f(t+x-x_ie_i)+f(t+x-2t_ie_i-x_ie_i)}{2}dt. \tag{4.8}$$

In view of Statements 2) and 3) of Lemma 4.4.1, from the equality (4.8) we obtain

$$\frac{\partial^2 U(f;x-x_ie_i,x_{k+1})}{\partial x_i^2}-\mathbf{D}_{x_i(x_{M|i})}f(0)$$

$$=\frac{1}{2}C_kx_{k+1}\int\limits_{R^k}\frac{[(k+3)t_i^2-|t|^2-x_{k+1}^2]t_i^2}{(|t|^2+x_{k+1}^2)^{\frac{k+5}{2}}}$$

$$\times\Big[\frac{f(t+x-x_ie_i)+f(t+x-2t_ie_i-x_ie_i)-2f(t-t_ie_i+x-x_ie_i)}{t_i^2}$$

$$-\mathbf{D}^2_{x_i(x_{M|i})}f(0)\Big]dt.$$

With Statements 4) and 6) of Lemma 4.4.1 taken into account, from the last equality we conclude that Item (a) is valid.

Proof of Item (b). The proof is carried out for the cases $i=1$ and $k=2$. We have

$$\frac{\partial^2 U(f;x_1,x_2,x_3)}{\partial x_1^2}=\frac{3x_3}{2\pi}\int\limits_{R^2}\frac{4(t_1-x_1)^2-(t_2-x_2)^2-x_3^2}{(|t-x|^2+x_3^2)^{\frac{7}{2}}}f(t)dt$$

$$=\frac{3x_3}{2\pi}\int\limits_0^\infty\int\limits_0^{2\pi}\frac{4(\rho\cos\varphi-x_1)^2-(\rho\sin\varphi-x_2)^2-x_3^2}{[(\rho\cos\varphi-x_1)^2+(\rho\sin\varphi-x_2)^2+x_3^2]^{\frac{7}{2}}}f(\rho\cos\varphi,\rho\sin\varphi)\rho d\rho d\varphi.$$

Let $N(x_1, x_2, x_3) \to (0, 0, 0)$ so that $x_2 = 0$ and $x_3 = x_1$. Then

$$\frac{\partial^2 U(f; x_3, 0, x_3)}{\partial x_1^2}$$

$$= \frac{3x_3}{2\pi} \int_0^\infty \int_0^{2\pi} \frac{4\rho^2 \cos^2\varphi - 8\rho x_3 \cos\varphi - \rho^2 \sin^2\varphi + 3x_3^2}{(\rho^2 - 2\rho x_3 \cos\varphi + 2x_3^2)^{\frac{7}{2}}} f(\rho\cos\varphi, \rho\sin\varphi)\rho d\rho d\varphi.$$

For every x_1, x_2 and x_3, by Lemma 4.4.1, we have

$$\int_{R^2} \frac{4(t_1 - x_1)^2 - (t_2 - x_2)^2 - x_3^2}{(|t - x|^2 + x_3^2)^{\frac{7}{2}}} dt_1 dt_2 = 0.$$

This, in particular, implies

$$\int_0^\infty \int_0^{2\pi} \frac{4\rho^2 \cos^2\varphi - 8\rho x_3 \cos\varphi - \rho^2 \sin\varphi + 3x_3^2}{(\rho^2 - 2\rho x_3 \cos\varphi + 2x_3^2)^{\frac{7}{2}}}$$

$$= 2\int_0^\infty \int_0^{\pi} \frac{4\rho^2 \cos^2\varphi - 8\rho x_3 \cos\varphi - \rho^2 \sin^2\varphi + 3x_3^2}{(\rho^2 - 2\rho x_3 \cos\varphi + 2x_3^2)^{\frac{7}{2}}} \rho d\rho d\varphi = 0.$$

Hence

$$\int_0^\infty \int_0^{\pi} \frac{4\rho^2 \cos^2\varphi - 8\rho x_3 \cos\varphi - \rho^2 \sin^2\varphi + 3x_3^2}{(\rho^2 - 2\rho x_3 \cos\varphi + 2x_3^2)^{\frac{7}{2}}} \rho d\rho d\varphi = 0. \tag{4.9}$$

In the interval $\frac{\pi}{2} \le \varphi \le \pi$, we have

$$x_3 \int_0^\infty \int_{\frac{\pi}{2}}^{\pi} \frac{4\rho^2 \cos^2\varphi - 8\rho x_3 \cos\varphi - \rho^2 \sin^2\varphi + 3x_3^2}{(\rho^2 - 2\rho x_3 \cos\varphi + 2x_3^2)^{\frac{7}{2}}} \rho d\rho d\varphi$$

$$> x_3 \int_0^\infty \int_{\frac{\pi}{2}}^{\pi} \frac{5\rho^2 \cos^2\varphi - \rho^2}{(\rho^2 + 4\rho x_3 + 4x_3^2)^{\frac{7}{2}}} \rho d\rho d\varphi = x_3 \int_0^\infty \int_{\frac{\pi}{2}}^{\pi} \frac{5\rho^2 \cos^2\varphi - \rho^2}{(\rho + 2x_3)^7} \rho d\rho d\varphi$$

$$= \frac{3\pi}{4} x_3 \int_0^\infty \frac{\rho^3 d\rho}{(\rho + 2x_3)^7} > C x_3 \int_0^\infty \frac{\rho^3 d\rho}{(\rho + 2x_3)^7} = \frac{C}{x_3^2}.$$

Thus

$$\lim_{x_3 \to 0+} \int_0^\infty \int_{\frac{\pi}{2}}^{\pi} \frac{4\rho^2 \cos^2\varphi - 8\rho x_3 \cos\varphi - \rho^2 \sin^2\varphi + 3x_3^2}{(\rho^2 - 2\rho x_3 \cos\varphi + 2x_3^2)^{\frac{7}{2}}} \rho d\rho d\varphi = +\infty. \tag{4.10}$$

It follows from (4.9) and (4.10) that

$$\lim_{x_3\to 0+} x_3 \int_0^{\infty}\int_0^{\frac{\pi}{2}} \frac{4\rho^2\cos^2\varphi - 8\rho x_3\cos\varphi - \rho^2\sin^2\varphi + 3x_3^2}{(\rho^2 - 2\rho x_3\cos\varphi + 2x_3^2)^{\frac{7}{2}}}\rho d\rho d\varphi = -\infty. \tag{4.11}$$

We now define the function $f(t_1,t_2)$ as follows:

$$f(t_1,t_2) = \begin{cases} -1, & \text{when } \; t_1 > 0, \quad t_2 > 0, \\ 1, & \text{when } \; t_1 < 0, \quad t_2 > 0, \\ 0, & \text{when } \; -\infty < t_1 < \infty, \quad t_2 \le 0, \\ 0, & \text{when } \; t_1 = 0, \quad 0 < t_2 < \infty. \end{cases}$$

Clearly, for this function

$$\mathbf{D}^2_{x_1(x_2)} f(0,0) = \lim_{(t_1,x_2)\to(0,0)} \frac{f(t_1,x_2) + f(-t_1,x_2) - 2f(0,x_2)}{t_1^2} = 0.$$

However, as it follows from (4.10) and (4.11),

$$\frac{\partial^2 U(f;x_1,0,x_3)}{\partial x_1^2} \to +\infty,$$

as $(x_1,x_2,x_3)\to(0,0,0)$ along the chosen path.

The theorem is proved. □

Theorem 4.4.3. *If at the point x^0 there exists a finite derivative $\overline{f}''_{x_i(\overline{x}_{M|i})}(x^0)$, then*

$$\lim_{(x,x_{k+1})\xrightarrow{\wedge}(x^0,0)} \frac{\partial^2 U(f;x,x_{k+1})}{\partial x_i^2} - \overline{f}''_{x_i}(x^0) = \frac{\partial^2 f(x^0)}{\partial x_i^2}. \tag{4.12}$$

Proof. Let $x^0 = 0$. By Lemma 4.4.1, we have

$$\begin{aligned} &\frac{\partial^2 U(f;x,x_{k+1})}{\partial x_i^2} - \overline{f}''_{x_i(\overline{x}_{M|i})}(x^0) \\ &= C_k x_{k+1} \int_{R^k} \frac{(k+3)(t_i-x_i)^2 - |t-x|^2 - x_{k+1}^2}{(|t-x|^2 + x_{k+1}^2)^{\frac{k+5}{2}}} f(t)dt \\ &= C_k x_{k+1} \int_{R^k} \frac{(k+3)(t_i-x_i)^2 - |t-x|^2 - x_{k+1}^2}{(|t-x|^2 + x_{k+1}^2)^{\frac{k+5}{2}}} \cdot \frac{1}{2}t_i^2 \\ &\times \Big[\frac{f(t) - f(t-t_ie_i) - g(t-t_ie_i)t_i}{\frac{1}{2}t^2} - \overline{f}''_{x_i(\overline{x}_{M|i})}(0)\Big]dt, \end{aligned} \tag{4.13}$$

where $g(t) = f'_{t_i}(t)$ when $t\in V_\delta$ and, on CV_δ, is extended so that $g(t)\in L(R^k)$.

From (4.13), by virtue of Statement 6) of Lemma 4.4.1, we have

$$\frac{\partial^2 U(f;x,x_{k+1})}{\partial x_i^2} - \overline{f}''_{x_i(\overline{x}_{M|i})}(x^0)$$

$$= C_k x_{k+1} \int\limits_{R^k} \frac{(k+3)(t_i-x_i)^2 - |t-x|^2 - x_{k+1}^2}{\left(|t-x|^2 + x_{k+1}^2\right)^{\frac{k+5}{2}}} \cdot \frac{1}{2} t_i^2$$

$$\times \left[\frac{f(t) - f(t-t_ie_i) - f'_{i_1}(t-t_ie_i)t_i}{\frac{1}{2}t_i^2} - \overline{f}''_{x_i(\overline{x}_{M|i})}(0) \right] dt + o(1). \qquad (4.14)$$

Taking into account Statements 5) and 6) of Lemma 4.4.1, from (4.14) we obtain the equality (4.12).

Theorem 4.4.3 is proved. □

Analogously to Lemma 4.4.1, we prove the following

Lemma 4.4.2. *For every* $(x, x_{k+1}) \in R_+^{k+1}$ *the following statements* $(i, j = \overline{1,k}$, $i \neq j)$ *are valid:*

1) $\int\limits_{R^k} \frac{\partial^2 P(t-x, x_{k+1})}{\partial x_i \partial x_j} f(t - t_ie_i) t_v dt = 0,\ v = \overline{1,k};$

2) $\int\limits_{R^k} \frac{\partial^2 P(t-x, x_{k+1})}{\partial x_i \partial x_j} f(t - t_ie_i - t_je_j) t_v dt = 0,\ v = \overline{1,k};$

3) $\frac{\Gamma\left(\frac{k+1}{2}\right)}{\pi^{\frac{k+1}{2}}} \int\limits_{R^k} \frac{\partial^2 P(t-x, x_{k+1})}{\partial x_i \partial x_j} t_i t_j dt = 1;$

4) $\int\limits_{R^k} \left| \frac{\partial^2 P(t, x_{k+1})}{\partial t_i \partial t_j} \right| |t_r t_s| dt = O(1);\ r, s = \overline{1,k};$

5) $\int\limits_{R^k} \left| \frac{\partial^2 P(t-x, x_{k+1})}{\partial t_i \partial t_j} \right| |t_r t_s| dt = O(1);$ *for* $\frac{x_{k+1}}{|x-x^0|} \geq C > 0,\ r, s = \overline{1,k};$

6) $\lim\limits_{x_{k+1}\to 0} \sup\limits_{|t|\geq\delta>0} \left| \frac{\partial^2 P(t, x_{k+1})}{\partial t_i \partial t_j} \right| |t_r t_s| = 0;\ r, s = \overline{1,k};$

7) $\lim\limits_{x_{k+1}\to 0} \sup\limits_{|t|\geq\delta>0} \left| \frac{\partial^2 P(t, x_{k+1})}{\partial t_j^2} \right| |t_r t_s| = 0;\ r, s = \overline{1,k};$

Theorem 4.4.4.

(a) *If at the point* x^0 *there exists a finite derivative* $\mathbf{D}^2_{[x_ix_j](x)} f(x^0)$, $i \neq j$, *then*

$$\lim_{(x,x_{k+1})\to(x^0,0)} \frac{\partial^2 U(f;x,x_{k+1})}{\partial x_i \partial x_j} = \mathbf{D}^2_{x_ix_j} f(x^0).$$

(b) *There exists a continuous function* $f \in L(R^k)$ *such that for every* $B \subset M$, $m(B) < k$ *all derivatives* $\mathbf{D}^2_{[x_ix_j](\overline{x}_B)} f(0) = 0$, *but the limits*

$$\lim_{x_{k+1}\to 0+} \frac{\partial^2 U(f;0,x_{k+1})}{\partial x_i \partial x_j}$$

do not exist.

Proof of Item (a). Let $x^0 = 0$ and $B_k = \dfrac{(k+1)(k+3)\Gamma\left(\frac{k+1}{2}\right)}{\pi^{\frac{k+1}{2}}}$. It can be easily verified that

$$\frac{\partial^2 U(f;x,x_{k+1})}{\partial x_i \partial x_j} = B_k x_{k+1} \int\limits_{R^k} \frac{(t_i - x_i)(t_j - x_j) f(t) dt}{\left(|t-x|^2 + x_{k+1}^2\right)^{\frac{k+5}{2}}}$$
$$= B_k x_{k+1} \int\limits_{R^k} \frac{t_i t_j f(x+t) dt}{\left(|t|^2 + x_{k+1}^2\right)^{\frac{k+5}{2}}}.$$

By Statements 1), 2) and 3) of Lemma 4.4.2, we have

$$\frac{\partial^2 U(f;x,x_{k+1})}{\partial x_i \partial x_j} - \mathbf{D}^2_{[x_i x_j](x)} f(0) = B_k x_{k+1} \int\limits_{R^k} \frac{t_i^2 t_j^2}{\left(|t|^2 + x_{k+1}^2\right)^{\frac{k+5}{2}}}$$
$$\times \left[\frac{f(x+t) - f(x+t-t_i e_i) - f(x+t-t_j e_j) + f(x+t-t_i e_i - t_j e_j)}{t_i t_j} \right.$$
$$\left. - \mathbf{D}^2_{[x_i x_j](x)} f(0) \right] dt,$$

whence, taking into account Statements 4) and 6) of Lemma 4.4.2, it follows that Item (a) is valid.

Proof of Item (b). Let $k = 4$ and $\mathbf{D} = \left(\sum_{i=1}^4 t_i^2 \le 1,\ t_i \ge 0,\ i = \overline{1,4} \right)$. We define the function f as follows: $f(t) = \sqrt[5]{t_1 t_2 t_3 t_4}$ when $t \in \mathbf{D}$ and, on the set $R^4 | \mathbf{D}$, we extend it continuously so that $f \in L(R^4)$. It is not difficult to verify that for every $B \subset M$, $m(B) < 4$ we have $\mathbf{D}^2_{[x_i x_j](\overline{x}_B)} f(0) = 0$. But

$$\frac{\partial^2 U(f;0,x_5)}{\partial x_1 \partial x_2} = C x_5 \int\limits_{\mathbf{D}} \frac{t_1 t_2 \sqrt[5]{t_1 t_2 t_3 t_4}}{(|t|^2 + x_5^2)^{\frac{9}{2}}} dt + o(1)$$
$$= C x_5 \int\limits_0^1 \frac{\rho^{\frac{29}{5}} d\rho}{(\rho^2 + x_5^2)^{\frac{9}{2}}} + o(1) \to +\infty \quad \text{as} \quad x_5 \to 0+.$$

The theorem is proved. □

Corollary. *If f has a continuous derivative $f''_{x_i x_j}(x)$ at the point x^0, then*

$$\lim_{(x,x_{k+1}) \to (x^0,0)} \frac{\partial^2 U(f;x,x_{k+1})}{\partial x_i \partial x_j} = \frac{\partial^2 f(x^0)}{\partial x_i \partial x_j}.$$

From Corollary 4.4.2 of Theorem 4.4.1 and from Corollary of Theorem 4.4.4 it follows that the following theorem is valid.

Theorem 4.4.5. *If the function f is twice continuously differentiable at the point x^0, then*

$$\lim_{(x,x_{k+1})\to(x^0,0)} d_x^2 U(f;x,x_{k+1}) = d^2 f(x^0).$$

Theorem 4.4.6.
(a) *If at the point x^0 there exists a finite derivative $\mathbf{D}^2_{[x_i]x_j(x)} f(x^0) = 0$, then*

$$\lim_{(x,x_{k+1})\xrightarrow[x_j]{\wedge}(x^0,0)} \frac{\partial^2 U(f;x,x_{k+1})}{\partial x_i \partial x_j} = \mathbf{D}^2_{x_i x_j} f(x^0).$$

(b) *There exists a continuous function $f \in L(R^k)$ such that for every $B \subset M$, $m(B) < k-1$, all derivatives $\mathbf{D}^2_{[x_i]x_j(\overline{x}_B)]} f(0) = 0$, but the limits*

$$\lim_{x_k+1\to 0+} \frac{\partial^2 U(f;0,x_{k+1})}{\partial x_i \partial x_j}$$

do not exist.

Theorem 4.4.7.
(a) *If at the point x^0 there exists a finite derivative $\mathbf{D}^2_{x_i x_j(x_{M\setminus\{t,j\}})} f(x^0)$, then*

$$\lim_{(x,x_{k+1})\xrightarrow[x_i x_j]{\wedge}(x^0,0)} \frac{\partial^2 U(f;x,x_{k+1})}{\partial x_i \partial x_j} = \mathbf{D}_{x_i x_j} f(x^0).$$

(b) *There exists a continuous function $g \in L(R^k)$ such that for every $B \subset M$, $m(B) < k-2$ all derivatives $\mathbf{D}^2_{x_i x_j(\overline{x}_B)} g(0) = 0$, but the limits*

$$\lim_{x_k+1\to 0+} \frac{\partial^2 U(g;0,x_{k+1})}{\partial x_i \partial x_j}$$

do not exist.

Corollary. *If the function of two variables $z = f(x,y)$ has, at (x_0,y_0), a finite derivative*

$$\mathbf{D}^2 f(x_0,y_0) = \lim_{(t,\tau)\to(0,0)} \Big[\frac{f(x_0+t,y_0+\tau) - f(x_0+t,y_0) - f(x_0,y_0+t) + f(x_0,y_0)}{t\tau},$$

then

$$\lim_{(x,u,z)\xrightarrow{\wedge}(x_0,y_0,0)} \frac{\partial^2 U(f;x,y,z)}{\partial x \partial y} = \mathbf{D}^2 f(x_0,y_0).$$

Theorem 4.4.8. (a) *If f is a twice differentiable function at the point x^0, then*

$$\lim_{(x,x_{k+1})\overset{\wedge}{\longrightarrow}(x^0,0)} d_x^2U(f;x,x_{k+1})=d^2f(x^0).$$

(b) *There exists a continuous function $f\in L(R^k)$ such that it is differentiable at the point $x^0=(0,0)$ and has, at this point, all partial derivatives of any order, but the limits*

$$\lim_{x_{k+1}\to 0+}\frac{\partial^2U(f;0,x_{k+1})}{\partial x_i\partial x_j},\quad i,j=\overline{1,k},$$

do not exist.

Proof of Item (a). Let $x^0=0$. The validity of Item (a) follows from the equalities

$$\frac{\partial^2U(f;x,x_{k+1})}{\partial x_i^2}-\frac{\partial^2 f(0)}{\partial x_i^2}$$
$$=C_kx_{k+1}\int\limits_{R^k}\frac{[(k+3)(t_i-x_i)^2-|t-x|^2-x_{k+1}^2]|t|^2}{(|t-x|^2+x_{k+1}^2)^{\frac{k+5}{2}}}$$
$$\times\frac{f(t)-f(0)-\Big(\sum\limits_{v=1}^k t_v\frac{\partial}{\partial t_v}\Big)f(0)-\frac12\Big(\sum\limits_{v=1}^k t_v\frac{\partial}{\partial t_v}\Big)^2f(0)}{|t|^2}\,dt$$

and

$$\frac{\partial^2U(f;x,x_{k+1})}{\partial x_i\partial x_j}-\frac{\partial^2 f(0)}{\partial x_i\partial x_j}$$
$$=B_kx_{k+1}\int\limits_{R^k}\frac{(t_i-x_i)(t_j-x_j)|t|^2}{(|t-x|^2+x_{k+1}^2)^{\frac{k+5}{2}}}$$
$$\times\frac{f(t)-f(0)-\Big(\sum\limits_{v=1}^k t_v\frac{\partial}{\partial t_v}\Big)f(0)-\frac12\Big(\sum\limits_{v=1}^k t_v\frac{\partial}{\partial t_v}\Big)^2f(0)}{|t|^2}\,dt.$$

Proof of Item (b). Consider the function

$$f(t_1,t_2)=\begin{cases}\sqrt[3]{(2t_1-t_2)^2}, & \text{when}\quad (t_1,t_2)\in D=\{(t_1,t_2)\in R^2:\\ & 0\le t_1<\infty;\ \ \frac12t_1\le t_2\le 2t_1\},\\ 0, & \text{when}\quad (t_1,t_2)\in CD.\end{cases}$$

This function is continuous in R^2, differentiable at the point $(0,0)$ and has all partial derivatives of any order, equal to zero, but the limits

$$\frac{\partial^2U(f;0,0,x_3)}{\partial x_1\partial x_2}=\frac{15}{2\pi}x_3\int\limits_0^\infty dt_1\int\limits_{\frac12t_1}^{2t_1}\frac{t_1t_2\sqrt[3]{(2t_1-t_2)^2(t_2-\frac12t_1)^2}}{(t_1^2+t_2^2+x_3^2)^{\frac72}}\,dt_2$$

$$> Cx_3 \int\limits_{x_3}^{2x_3} t_1^2 dt_1 \frac{\sqrt[3]{(2t_1 - \frac{3}{2}t_1)^2(t_1 - \frac{1}{2}t_1)^2}}{(t_1^2 + 4t_1^2 + x_3^2)^{\frac{7}{2}}} dt_2$$

$$= \frac{C}{x_3^6} \int\limits_{x_3}^{2x_3} t_1^{\frac{13}{3}} dt_1 = \frac{C}{\sqrt[3]{x_3^2}} \to \infty \quad \text{as} \quad x_3 \to 0+$$

do not exist. □

Theorem 4.4.9.
(a) *If at the point x^0 there exists a finite derivative $\mathbf{D}^*_{x_i x_j(\overline{x}_{M|\{i,j\}})} f(x^0)$, then*

$$\lim_{x_{k+1} \to 0+} \frac{\partial^2 U(f; x^0, x_{k+1})}{\partial x_i \partial x_j} = \mathbf{D}^*_{x_i x_j} f(x^0).$$

(b) *There exists a continuous function $f \in L(R^k)$ such that $\mathbf{D}_{x_i x_j(\overline{x}_{M|\{i,j\}})} f(x^0) =$ 0, but the limits*

$$\lim_{(x, x_{k+1}) \xrightarrow{\wedge} (x^0, 0)} d_x^2 \frac{U(f; x, x_{k+1})}{\partial x_i \partial x_j}$$

do not exist.

Proof of Item (b). The proof is carried out for the case $k = 2$. Assume $D_1 = [0, 1; 0, 1]$, $D_2 = [-1, 0; 0, 1]$. Let

$$f(t_1, t_2) = \begin{cases} \sqrt{t_1 t_2}, & \text{when} \quad (t_1, t_2) \in D_1, \\ \sqrt{-t_1 t_2}, & \text{when} \quad (t_1, t_2) \in D_2, \\ 0, & \text{when} \quad t_2 \le 0, \end{cases}$$

and extend $f(t_1, t_2)$ continuously on the set $R_+^2 | (\mathbf{D}_1 \cup \mathbf{D}_2)$ so that $f \in L(R^2)$. It is not difficult to verify that $\mathbf{D}^* f(0, 0) = 0$. Let $x_1^0 = x_2^0 = 0$ and $(x_1, x_2, x_3) \to (0, 0, 0)$ so that $x_2 = 0$ and $x_3 = x_1$. Then for the constructed function,

$$\frac{\partial^2 U(f; x_1, x_2, x_3)}{\partial x_1 \partial x_2} = \frac{15 x_3}{2\pi} \left\{ \int\limits_0^1 \int\limits_0^1 \frac{(t_1 - x_1) t_2 \sqrt{t_1 t_2} dt_1 dt_2}{[(t_1 - x_1)^2 + t_2^2 + x_3^2]^{\frac{7}{2}}} \right.$$

$$\left. - \int\limits_{x_1}^{1+x_1} \int\limits_0^1 \frac{t_1 t_2 \sqrt{(t_1 - x_1) t_2} dt_1 t_2}{(t_1^2 + t_2^2 + x_1^2)^{\frac{7}{2}}} \right\} + o(1) = C x_1 \left\{ \int\limits_{-x_1}^{x_1} \int\limits_0^1 \frac{t_1 t_2 \sqrt{t_2 (t_1 + x_1)} dt_1 t_2}{(t_1^2 + t_2^2 + x_1^2)^{\frac{7}{2}}} \right.$$

$$\left. + \int\limits_{-1}^0 \int\limits_0^1 \frac{(t_1 - x_1) t_2 \sqrt{-t_1 t_2} dt_1 dt_2}{[(t_1 - x_1)^2 + t_2^2 + x_3^2]^{\frac{7}{2}}} \right\} + o(1) = C x_1 \left\{ \int\limits_{-x_1}^{x_1} \int\limits_0^1 \frac{t_1 t_2 \sqrt{t_2 (t_1 + x_1)} dt_1 dt_2}{(t_1^2 + t_2^2 + x_1^2)^{\frac{7}{2}}} \right.$$

$$+\int_{x_1}^{1-x_1}\int_0^1 \frac{t_1t_2[\sqrt{t_2(x_1+t_1)}-\sqrt{t_2(t_1-x_1)}]dt_1dt_2}{[t_1^2+t_2^2+x_1^2]^{\frac{7}{2}}}$$

$$-\int_{1-x_1}^{1+x_1}\int_0^1 \frac{t_1t_2\sqrt{(t_1-x_1)t_2}dt_1dt_2}{[t_1^2+t_2^2+x_1^2]^{\frac{7}{2}}}\Bigg\}+o(1)=Cx_1(I_1+I_2-I_3)+o(1).$$

It is easy to show that

$$I_1=\int_0^{x_1}\int_0^1 \frac{t_1t_2[\sqrt{t_2(t_1+x_1)}-\sqrt{t_2(x_1-t_1)}]}{(t_1^2+t_2^2+x_1^2)^{\frac{7}{2}}}dt_1dt_2>0, \quad I_3=O(1). \tag{4.15}$$

Furthermore,

$$I_2>\int_{x_1}^{2x_1}\int_0^{x_1} \frac{t_1t_2[\sqrt{t_2(x_1+t_1)}-\sqrt{t_2x_1}]}{(t_1^2+t_2^2+x_1^2)^{\frac{7}{2}}}dt_1dt_2$$

$$=\int_0^{x_1} t_2^{\frac{3}{2}}dt_2\int_{x_1}^{2x_1}\frac{t_1(\sqrt{x_1+t_1}-\sqrt{x_1})}{(t_1^2+t_2^2+x_1^2)^{\frac{7}{2}}}dt_1$$

$$>\int_0^{x_1} t_2^{\frac{3}{2}}dt_2\int_{x_1}^{2x_1}\frac{x_1\left(\sqrt{2x_1}-\sqrt{x_1}\right)}{(4x_1^2+x_1^2+x_1^2)^{\frac{7}{2}}}dt_1>\frac{C}{x_1^2}. \tag{4.16}$$

Consequently, for $x_2=0$ and $x_3=x_1$, from (4.15) and (4.16), we have

$$\frac{\partial^2U(f;x_1,0,x_1)}{\partial x_1\partial x_2}>\frac{C}{x_1},$$

whence

$$\frac{\partial^2U(f;x_1,x_2,x_3)}{\partial x_1\partial x_2}\to+\infty$$

as $(x_1,x_2,x_3)\to(0,0,0)$ along the chosen path. □

Theorem 4.4.10. *If at the point x^0 there exists a finite derivative $\mathbf{D}^*_{[x_ix_j](x)}f(x^0)$, then*

$$\lim_{(x,x_{k+1})\to(x^0,0)}\frac{\partial^2U(f;x,x_{k+1})}{\partial x_i\partial x_j}=\mathbf{D}^*_{x_ix_j}f(x^0).$$

Item (a) of Theorem 4.4.4 is a corollary of Theorem 4.4.10.

Theorem 4.4.11. *If at the point x^0 there exists a finite derivative $\mathbf{D}^*_{[x_i]x_j(x)}f(x^0)$, then*

$$\lim_{(x-x_ie_i-x_j^0e_j,x_{k+1})\to(x^0,0)}\frac{\partial^2U(f,x-x_ie_i-x_j^0e_j,x_{k+1})}{\partial x_i\partial x_j}=\mathbf{D}^*_{x_ix_j}f(x^0).$$

4.5 The Dirichlet Problem for a Half-Space R_+^3

The Dirichlet problem for the Laplace equation consists in finding such a function $U(x, y, z)$ in the domain R_+^3 with the boundary R^2 that satisfying in that domain the equation $\Delta U = 0$ and the boundary condition $U|_{R^2} = f(x, y)$. The solution of the problem is given in [61] for the case $f \in L(R^2)$. In this section this problem is solved for the case in which the boundary function is measurable and finite almost everywhere on R^2, i.e., in N.N. Luzin's formulation (see Theorem 4.5.2).

Theorem 4.5.1. *Let f be an arbitrary measurable and finite function almost everywhere on R^2. Then there exists a bounded continuous function F such that*

$$\mathbf{D}^2 F(x, y) = f(x, y)$$

almost everywhere on R^2.

Proof. By T_n we denote the square $[-n, n; -n; n]$. ∂T_n is the contour of the square. By the well-known theorem ([57], p. 314), there exists in T_1 a continuous function Φ_1 such that

$$\mathbf{D}^2 \Phi_1(x, y) = f(x, y)$$

almost everywhere in T_1. Let $M = \text{const} > 0$. We construct in T_1 a continuous step function S_1 such that ([14], p. 16)

$$\begin{aligned} S_1(x, y) &= \Phi_1(x, y), \quad \text{when} \quad (x, y) \in \partial T_1, \\ |S_1(x, y) - \Phi_1(x, y)| &\le M, \quad \text{for} \quad (x, y) \in T_1. \end{aligned}$$

Assume

$$F_1(x, y) = \begin{cases} \Phi_1(x, y) - S_1(x, y), & \text{for} \quad (x, y) \in T_1, \\ 0, & \text{for} \quad (x, y) \in CT_1. \end{cases}$$

Clearly, $F_1(x, y) = 0$ when $(x, y) \in \partial T_1$, F_1 is continuous in R^2, $|F_1(x, y)| \le M$ for all $(x, y) \in R^2$ and

$$D^2 F_1(x, y) = \begin{cases} f(x, y) & \text{almost everywhere in} \quad T_1, \\ 0 & \text{almost everywhere in} \quad CT_1. \end{cases}$$

We construct a continuous function F_2 in R^2 so that $F_2(x, y) = 0$ for $(x, y) \in CT_2 \cup T_1$, $|F_2(x, y)| \le M$ for all $(x, y) \in R^2$, and

$$D^2 F_2(x, y) = \begin{cases} f(x, y) & \text{almost everywhere in} \quad T_2 - T_1, \\ 0 & \text{almost everywhere in} \quad CT_2 \cup T_1. \end{cases}$$

Continuing this process, we obtain a sequence of continuous functions $\{F_n\}$ in R^2 such that $F_{n+1}(x, y) = 0$ for $(x, y) \in CT_{n+1} \cup T_n$, $|F_{n+1}(x, y)| \le M$ for all

$(x,y)\in R^2$ and

$$D^2F_{n+1}(x,y)=\begin{cases} f(x,y) & \text{almost everywhere in} \quad T_{n+1}-T_n,\\ 0 & \text{almost everywhere in} \quad CT_{n+1}\cup T_n. \end{cases}$$

Assume

$$F(x,y)=\sum_{k=1}^{\infty}F_k(x,y).$$

It is evident that F is continuous in R^2, $|F(x,y)|\le M$ for all $(x,y)\in R^2$, and

$$\mathbf{D}^2F(x,y)=f(x,y)$$

almost everywhere in R^2.

Consequently, F is the required function. □

Theorem 4.5.2. *Let f be an arbitrary measurable and finite function almost everywhere on R^2. Then there exists a harmonic function U in R^3_+ such that*

$$\lim_{(x,y,z)\overset{\wedge}{\longrightarrow}(x_0,y_0,0)} U(x,y,z)=f(x_0,y_0)$$

almost everywhere on R^2.

Proof. By Theorem 4.5.1, for f we construct, in R^2, a continuous bounded function F such that the equality

$$\mathbf{D}^2F(x,y)=f(x,y)$$

is fulfilled almost everywhere on R^2.

Consider the expression

$$\begin{aligned} U(x,y,z)&=\frac{z}{2\pi}\iint\limits_{R^2}F(t,\tau)\frac{\partial^2}{\partial t\partial\tau}\Big\{\frac{1}{[(x-t)^2+(y-\tau)^2+z^2]^{\frac{3}{2}}}\Big\}dtd\tau\\ &=\frac{15z}{2\pi}\iint\limits_{R^2}\frac{(t-x)(\tau-y)}{[(t-x)^2+(\tau-y)^2+z^2]^{\frac{7}{2}}}F(t,\tau)dtd\tau. \end{aligned}$$

It is not difficult to show that U is a harmonic function in R^3_+. By Corollary of Theorem 4.4.7, if at the point $(x_0,y_0,0)$ there exists $\mathbf{D}^2F(x_0,y_0)$, then

$$\lim_{(x,y,z)\overset{\wedge}{\longrightarrow}(x_0,y_0,0)} U(x,y,z)=\mathbf{D}^2F(x_0,y_0).$$

Since $\mathbf{D}^2F(x,y)=f(x,y)$ almost everywhere, the theorem is complete. □

4.6 Generalized Partial Derivatives of Arbitrary Order

We use the same notation as in Sections 4.1–4.3.

Below it will be assumed that the considered functions belong to the space $\widetilde{L}(R^k)$. Let $u \in R$. We will deal with the following generalized partial derivatives of arbitrary order of the function $f(x) = f(x_1, x_2, \dots, x_k)$:

1) If there exist functions $a_i(\overline{x}_B)$ (if $B = \varnothing$, then $a_i(\overline{x}_B) = a_i = \text{const}$), $i = \overline{0, r-1}$ and a number a_r such that there exist limits $\lim_{\overline{x}_B \to \overline{x}^0_B} a_i(\overline{x}_B) = a_i$ and in the neighborhood of the point x_0 we have

$$f(x_B + x^0_{B'} + ue_i) = a_0(\overline{x}_B) + a_1(\overline{x}_B)\frac{u}{1!} + a_2(\overline{x}_B)\frac{u^2}{2!} + \cdots$$
$$\cdots + a_{r-1}(\overline{x}_B)\frac{u^{r-1}}{(r-1)!} + (a_r + \varepsilon(u, \overline{x}_B)\frac{u^r}{r!},$$

where $\lim\limits_{\substack{u\to 0,\\ \overline{x}_B \to \overline{x}^0_B}} \varepsilon(u, \overline{x}_B) = 0$, then at the point x^0 the function $f(x)$ has, with respect to the variable x_i, the generalized r th partial derivative equal to a_r and we denote it

a) by $\mathbf{D}^{(r)}_{x_i} f(x^0)$ if $B = \varnothing$,

b) by $\mathbf{D}^{(r)}_{x_i(\overline{B})} f(x^0)$ if $i \in B'$,

c) by $\overline{\mathbf{D}}^{(r)}_{x_i(\overline{B})} f(x^0)$ if $i \in B$.

From the definition it follows that if for some $B \subset M$ there exists $\overline{\mathbf{D}}^{(r)}_{x_i(\overline{B})} f(x^0)$, then there exist $\mathbf{D}^{(r)}_{x_i(\overline{B}|i)} f(x^0)$ and

$$\overline{\mathbf{D}}^{(r)}_{x_i(\overline{B})} f(x^0) = \mathbf{D}^{(r)}_{x_i(\overline{B}|i)} f(x^0) = \mathbf{D}^{(r)}_{x_i} f(x^0).$$

2) Let r be an odd number. If there exist functions $b_{2i-1}(\overline{x}_B)$, $i = 1, 2, \dots, \dfrac{(r-1)}{2}$ and a number b_r such that there exist limits $\lim\limits_{\overline{x}_B \to \overline{x}^0_B} b_{2i-1}(\overline{x}_B) = b_{2i-1}$ and in the neighborhood of the point x^0

$$\frac{1}{2}(f(x_B + x^0_{B'} + ue_i) - f(x_B + x^0_{B'} - ue_i))$$
$$= \sum_{v=1}^{\frac{(r-1)}{2}} b_{2v-1}(\overline{x}_B)\frac{u^{2v-1}}{(2v-1)!} + (b_r + \varepsilon(u, \overline{x}_B))\frac{u^r}{r!},$$

where $\lim\limits_{\substack{u\to 0,\\ \overline{x}_B \to \overline{x}^0_B}} \varepsilon(u, \overline{x}_B) = 0$, then at the point x^0 the function $f(x)$ has, with respect to the variable x_i, the generalized partial r th symmetric derivative equal to b_r, which we denote

a) by $\mathbf{D}^{*(r)}_{x_i} f(x^0)$ if $B = \varnothing$,

b) by $\mathbf{D}^{*(r)}_{x_i(\overline{x}_B)} f(x^0)$ if $i \in B'$,

c) by $\overline{\mathbf{D}}^{*(r)}_{x_i(\overline{x}_B)} f(x^0)$ if $i \in B$.

The same definition is valid when r is even, but in that case the difference $f(x_B + x^0_{B'} + ue_i) - f(x_B + x^0_{B'} - ue_i)$ should be replaced by the sum $f(x_B + x^0_{B'} + ue_i) + f(x_B + x^0_{B'} - ue_i)$.

It is easy to verify that from the existence of the derivatives $\mathbf{D}^{(r)}_{x_i} f(x^0)$ follows the existence of derivatives $\mathbf{D}^{*(r)}_{x_i} f(x^0)$ and their equality.

For symmetric derivatives, from the existence of $\mathbf{D}^{*(r)}_{x_i} f(x^0)$ follows the existence of $\mathbf{D}^{*(r-2)}_{x_i} f(x^0)$ though $\mathbf{D}^{*(r-1)}_{x_i} f(x^0)$ may not exist ([35], p. 93).

4.7 The Boundary Properties of Arbitrary Order Partial Derivatives of the Poisson Integral for a Half-Space R^{k+1}_+

In this section we prove the Fatou type theorems on the boundary properties of arbitrary order partial derivatives of the Poisson integral for a half-space R^{k+1}_+ ([107]).

Lemma 4.7.1. *For every $r \in N$ and $(x, x_{k+1}) \in R^{k+1}_+$ the following statements are valid:*

1) $J^{(r)}_v = \int\limits_{-\infty}^{\infty} \dfrac{\partial^r P(t - x, x_{k+1})}{\partial x_i^r}\, t_i^v dt_i = 0,\ i = \overline{1,k},\ v = \overline{0, r-1};$

2) $J_r = \dfrac{\Gamma\left(\frac{k+1}{2}\right)}{\pi^{\frac{k+1}{2}}} \int\limits_{R^k} \dfrac{\partial^r P(t - x, x_{k+1})}{\partial x_i^r} \dfrac{t_i^r}{r!}\, dt = 1;$

3) $\int\limits_{R_k} \left| \dfrac{\partial^r P(t, x_{k+1})}{\partial t_i^r} \right| |t|^r dt < C;$

4) $\int\limits_{R^k} \left| \dfrac{\partial^r P(t - x, x_{k+1})}{\partial t_i^r} \right| |t|^r dt < C$ *for* $\dfrac{x_{k+1}}{|x_i|} \ge C > 0;$

5) $\sup\limits_{|t| \ge \delta > 0} \left| \dfrac{\partial^r P(t, x_{k+1})}{\partial t_i^r} \right| (|t|^2 + x^2_{k+1})^{\frac{k+1}{2}} |t|^v < C x_{k+1},\ \ v = \overline{0, r}.$

Proof. Statement 1) is proved by induction. When $r = 1$ and $r = 2$, the validity of the statement is shown in Sections 4.2 and 4.3.

Let us assume now that for $r = n$ the the equalities

$$J^{(n)}_v = \int\limits_{-\infty}^{\infty} \frac{\partial^n P(t - x, x_{k+1})}{\partial x_i^n}\, t_i^v dt_i = 0, \quad v = \overline{0, n-1},$$

are fulfilled and show that the equality

$$J_v^{(n+1)} = \int_{-\infty}^{\infty} \frac{\partial^{n+1} P(t-x, x_{k+1})}{\partial x_i^{n+1}} t_i^v dt_i = 0, \quad v = \overline{0, n},$$

holds.

Indeed, using integration by parts, we obtain

$$0 = J_v^{(n)} = (-1)^n \int_{-\infty}^{\infty} \frac{\partial^n P(t-x, x_{k+1})}{\partial t_i^n} t_i^v dt_i$$

$$= \frac{(-1)^{n+1}}{v+1} \int_{-\infty}^{\infty} \frac{\partial^{n+1} P(t-x, x_{k+1})}{\partial t_i^{n+1}} t_i^{v+1} dt_i = \frac{1}{v+1} \int_{-\infty}^{\infty} \frac{\partial^{n+1} P(t-x, x_{k+1})}{\partial x_i^{n+1}} t_i^{v+1} dt_i.$$

Hence $J_v^{(n+1)} = 0$ when $v = \overline{1, n}$, and for $v = 0$ we likewise have

$$J_0^{(n+1)} = \frac{\partial^{n+1}}{\partial x_i^{n+1}} \int_{-\infty}^{\infty} P(t-x, x_{k+1}) dt = 0.$$

The validity of Statement 2) is also proved by induction. For $r = 1$ and $r = 2$, the validity of the statement is shown in Sections 4.2 and 4.3.

Assume now that for $r = n$ the equality

$$J_n = \frac{\Gamma\left(\frac{k+1}{2}\right)}{\pi^{\frac{k+1}{2}}} \int_{R^k} \frac{\partial^n P(t-x, x_{k+1})}{\partial x_i^n} \frac{t_i^n}{n!} dt = 1$$

is fulfilled. Then we can prove that the equality

$$J_{n+1} = \frac{\Gamma\left(\frac{k+1}{2}\right)}{\pi^{\frac{k+1}{2}}} \int_{R^k} \frac{\partial^{n+1} P(t-x, x_{k+1})}{\partial x_i^{n+1}} \frac{t_i^{n+1}}{(n+1)!} dt = 1$$

is valid.

Indeed, using integration by parts, we obtain

$$\int_{-\infty}^{\infty} \frac{\partial^n P(t-x, x_{k+1})}{\partial x_i^n} \frac{t_i^n}{n!} dt_i = . \int_{-\infty}^{\infty} \frac{\partial^{n+1} P(t-x, x_{k+1})}{\partial x_i^{n+1}} \frac{t_i^{n+1}}{(n+1)!} dt_i.$$

Taking this into account, we have

$$1 = J_n = \frac{\Gamma\left(\frac{k+1}{2}\right)}{\pi^{\frac{k+1}{2}}} \int_{R^{k-1}} d\bar{t}_{M|i} \int_{-\infty}^{\infty} \frac{\partial^{n+1} P(t-x, x_{k+1})}{\partial x_i^{n+1}} \frac{t_i^{n+1}}{(n+1)!} dt_i$$

$$= \frac{\Gamma\left(\frac{k+1}{2}\right)}{\pi^{\frac{k+1}{2}}} \int\limits_{R^k} \frac{\partial^{n+1} P(t-x, x_{k+1})}{\partial x_i^{n+1}} \frac{t_i^{n+1}}{(n+1)!} dt = J_{n+1}.$$

Let us now prove the validity of Statement 3). We have

$$|t|^r \frac{\partial^r P(t, x_{k+1})}{\partial t_i^r} = |t|^r x_{k+1} \frac{\partial^r}{\partial t_i^r} \left(\frac{1}{\left(|t|^2 + x_{k+1}^2\right)^{\frac{k+1}{2}}} \right)$$
$$= |t_i|^r x_{k+1} \frac{I(t_1, t_2, \dots, t_k, x_{k+1})}{\left(|t|^2 + x_{k+1}^2\right)^{\frac{k+2r+1}{2}}},$$

where $I(t_1, t_2, \dots, t_k, x_{k+1})$ is a homogeneous polynomial of degree r of $(t_1, t_2, \dots, t_k, x_{k+1})$.

Passing to the spherical coordinates, we obtain

$$\int\limits_{R^k} \left| \frac{\partial^r P(t, x_{k+1})}{\partial t_i^r} \right| |t|^r dt \leq C x_{x+1} \int\limits_0^\infty \frac{T(\rho, x_{k+1}) \rho^{r+k-1}}{\left(\rho^2 + x_{k+1}^2\right)^{\frac{k+2r+1}{2}}} d\rho,$$

where $T(\rho, x_{k+1}) > 0$ is a homogeneous polynomial of degree r in (ρ, x_{k+1}). Using the substitution $\rho = x_{k+1}\rho_1$, we obtain

$$\int\limits_{R^k} \left| \frac{\partial^r P(t, x_{k+1})}{\partial t_i^r} \right| |t|^r dt < C \sum_{\nu=0}^{r} \int\limits_0^\infty \frac{\rho_1^{k+r+v-1} d\rho_1}{\left(1 + \rho_1^2\right)^{\frac{k+2r+1}{2}}} = O(1).$$

Statement 4) follows from Statement 3) if we take into account the conditions

$$\frac{x_{k+1}}{|x_i|} \geq C > 0.$$

Statement 5) follows from the inequality

$$\left| \frac{\partial^r P(t, x_{k+1})}{\partial t_i^r} \right| \left(|t|^2 + x_{k+1}^2\right)^{\frac{k+1}{2}} |t|^v < x_{k+1} \frac{|I(t_1, t_2, \dots, t_k, x_{k+1})|}{\left(|t|^2 + x_{k+1}^2\right)^{r - \frac{v}{2}}}, \quad v = \overline{0, r}$$

since $I(t_1, t_2, \dots, t_k, x_{k+1})$ is a homogeneous polynomial of degree r.

Lemma 4.7.1. is proved. □

Theorem 4.7.1. *If at the point x^0 there exists a finite derivative $\mathbf{D}^{(r)}_{x_i(\overline{x}_{M|i})} f(x^0)$, then*

$$\lim_{(x, x_{k+1}) \xrightarrow[x_i]{\wedge} (x^0, 0)} \frac{\partial^r U(f; x, x_{k+1})}{\partial x_i^r} = \mathbf{D}^{(r)}_{x_i} f(x^0).$$

Proof. Let $x^0=0$. By Statements 1) and 2), from Lemma 4.7.1 we have

$$\frac{\partial^r U(f;x,x_{k+1})}{\partial x_i^r}=\frac{\Gamma\left(\frac{k+1}{2}\right)}{\pi^{\frac{k+1}{2}}}\int\limits_{R^k}\frac{\partial^r P(t-x,x_{k+1})}{\partial x_i^r}$$

$$\times\left(\frac{r!}{t_i^r}\left(f(t)-\sum_{v=0}^{r-1}a_v(\overline{t}_{M|i})\frac{t_i^v}{v!}\right)-\mathbf{D}^{(r)}_{x_i(\overline{x}_{M|i})}f(0)\right)\frac{t_i^r}{r!}dt+\mathbf{D}^{(r)}_{x_i(\overline{x}_{M|i})}f(0)$$

$$=C\Big(\int\limits_{V_\delta}+\int\limits_{CV\delta}\Big)+\mathbf{D}^{(r)}_{x_i(\overline{x}_{M|i})}f(0)=C(J_1+J_2)+\mathbf{D}^{(r)}_{x_i(\overline{x}_{M|i})}f(0),$$

where V_δ is a ball with center at the point 0 and of radius δ. Let $\varepsilon>0$. We choose $\delta>0$ such that

$$\left|\frac{r!}{t_i^r}\left(f(t)-\sum_{v=0}^{r-1}a_v(\overline{t}_{M|i})\frac{t_i^\nu}{\nu!}\right)-\mathbf{D}^{(r)}_{x_i(\overline{x}_{M|i})}f(0)\right|<\varepsilon,$$

for $|t|<\delta$. By virtue of this inequality and Statement 4) of Lemma 4.7.1, we have

$$|J_1|<C\varepsilon\quad\text{for}\quad\frac{x_{k+1}}{|x_i|}\geq C>0. \tag{7.1}$$

Further, taking into account this inequality and Statement 5) of Lemma 4.7.1, we get

$$|J_2|<Cx_{k+1}\quad\text{for}\quad|x|<\frac{\delta}{2}. \tag{7.2}$$

From (7.1) and (7.2) (assuming that ε is arbitrarily small) it follows that Theorem 4.7.1 is valid. □

Theorem 4.7.2.
(a) *If at the point x^0 there exists a finite derivative $\mathbf{D}^{*(r)}_{x_i(\overline{x}_{M|i})}f(x^0)$, then*

$$\lim_{(x-x_ie_i+x_i^0e_i,x_{k+1})\to(x^0,0)}\frac{\partial^r U(f;x-x_ie_i+x_i^0e_i,x_{k+1})}{\partial x_i^r}=\mathbf{D}^{*(r)}_{x_i}f(x^0).$$

(b) *There exist functions φ and g such that $\mathbf{D}^{*(1)}_{x_i(\overline{x}_{M|i})}\varphi(x^0)$ and $\mathbf{D}^{*(2)}_{x_i(\overline{x}_{M|i})}g(x^0)$ exist, however, there are no limits*

$$\lim_{(x,x_k+1)\xrightarrow{\wedge}(x^0,0)}\frac{\partial U(\varphi;x,x_{k+1})}{\partial x_i},\qquad\lim_{(x,x_k+1)\xrightarrow{\wedge}(x^0,0)}\frac{\partial^2 U(g;x,x_{k+1})}{\partial x_i^2}.$$

Proof of Item (a). Let $x^0 = 0$, and let r be an even number. We have

$$\frac{\partial^r U(f; x - x_i e_i, x_{k+1})}{\partial x_i^r} = \frac{\Gamma\left(\frac{k+1}{2}\right)}{\pi^{\frac{k+1}{2}}} \int\limits_{R^k} \frac{\partial^r P(t - x + x_i e_i, x_{k+1})}{\partial x_i^r} f(t)dt$$

$$= \frac{\Gamma\left(\frac{k+1}{2}\right)}{\pi^{\frac{k+1}{2}}} \int\limits_{R^k} \frac{\partial^r P(t, x_{k+1})}{\partial x_i^r} f(x + t - x_i e_i)dt.$$

Note that for even r $\partial^r P(t - x + x_i e_i, x_{k+1})/\partial x_i^r$ is an even function of t_i. Therefore, using first the substitution $t_i = -\tau_i$ and then $t = \tau + x - x_i e_i$, we find that

$$\frac{\partial^r U(f; x - x_i e_i, x_{k+1})}{\partial x_i^r} = \frac{\Gamma\left(\frac{k+1}{2}\right)}{\pi^{\frac{k+1}{2}}} \int\limits_{R^k} \frac{\partial^r P(t - x + x_i e_i, x_{k+1})}{\partial x_i^r} f(t - 2t_1 e_i)dt$$

$$= \frac{\Gamma\left(\frac{k+1}{2}\right)}{\pi^{\frac{k+1}{2}}} \int\limits_{R^k} \frac{\partial^r P(t, x_{k+1})}{\partial x_i^r} f(x + t - x_i e_i - 2t_i e_i)dt,$$

whence we obtain

$$\frac{\partial^r U(f; x - x_i e_i, x_{k+1})}{\partial x_i^r}$$

$$= \frac{\Gamma\left(\frac{k+1}{2}\right)}{\pi^{\frac{k+1}{2}}} \int\limits_{R^k} \frac{\partial^r P(t, x_{k+1})}{\partial x_i^r} \frac{f(x + t - x_i e_i) + f(x + t - x_i e_i - 2t_i e_i)}{2} dt.$$

By Statements 1) and 2) of Lemma 4.7.1, we have

$$\frac{\partial^r U(f; x - x_i e_i, x_{k+1})}{\partial x_i^r}$$

$$= \frac{\Gamma\left(\frac{k+1}{2}\right)}{\pi^{\frac{k+1}{2}}} \int\limits_{R^k} \frac{\partial^r P(t, x_{k+1})}{\partial x_i^r} \left(\frac{r!}{t_i^r} \left(\frac{f(x + t - x_i e_i) + f(x + t - x_i e_i - 2t_i e_i)}{2} \right.\right.$$

$$\left.\left. - \sum_{v=0}^{\frac{r-2}{2}} b_{2v}(\overline{x}_{M|i}) \frac{t^{2v}}{(2v)!} \right) - \mathbf{D}^{*(r)}_{x_i(\overline{x}_{M|i})} f(0) \right) \frac{t_i^r}{r!} dt + \mathbf{D}^{*(r)}_{x_i(\overline{x}_{M|i})} f(0).$$

Hence, by virtue of Statements 3) and 5) of Lemma 4.7.1, we obtain the validity of Item (a).

The validity of Item (b) is proved in Theorems 4.2.5. and 4.4.2. □

Theorem 4.7.3. (a) *If at the point* x^0 *there exists a finite derivative* $\overline{\mathbf{D}}^{*(r)}_{x_i(x)} f(0)$, *then*

$$\lim_{(x,x_{k+1})\to(x^0,0)} \frac{\partial^r U(f;x,x_{k+1})}{\partial x_i^r} = \mathbf{D}^{*(r)}_{x_i} f(x^0).$$

(b) *There exist continuous functions* φ *and* g *such that for every* $B \subset M$, $m(B) < k$ *all derivatives* $\overline{\mathbf{D}}^{*(1)}_{x_i(\overline{x}_B)} \varphi(x^0)$ *and* $\overline{\mathbf{D}}^{*(2)}_{x_i(\overline{x}_B)} g(x^0)$, $i = \overline{1,k}$ *exist, however, there are no limits*

$$\lim_{x_{k+1}\to 0+} \frac{\partial U(\varphi;x^0,x_{k+1})}{\partial x_i}, \qquad \lim_{x_{k+1}\to 0+} \frac{\partial^2 U(g;x^0,x_{k+1})}{\partial x_i^2}.$$

Item (a) can be proved analogously to Item (a) of Theorem 4.7.2. Item (b) is proved in Theorems 4.2.1 and 4.4.1.

4.8 Generalized Mixed Derivatives and Differentials of Arbitrary Order

We use the same notation as in Sections 4.1–4.3, $\alpha = (\alpha_1, \alpha_2, \dots, \alpha_s)$ and $\beta = (\beta_1, \beta_2 \dots, \beta_s)$ are multi-indices, α_i and β_i are nonnegative integers, $|\alpha| = \alpha_1 + \alpha_2 + \cdot + \alpha_s$; $\alpha! = \alpha_1!\alpha_2! \dots \alpha_s!$; if $B = \{i_1, i_2, \dots, i_S\} \subset M = \{1, 2, \dots, k\}$, $1 \le s \le k$ ($i_l < i_r$, for $l < r$), then $\overline{x}_B = (x_{i_1}, x_{i_2}, \dots, x_{i_s}) \in R^s$; $\overline{x}_B^\alpha = x_{i_1}^{\alpha_1} x_{i_2}^{\alpha_2} \dots x_{i_s}^{\alpha_s}$; $m(\alpha)$ is a number of coordinates for α; $\mathbf{D}^{(\alpha)} = \left(\frac{\partial}{\partial x_1}\right)^{\alpha_1} \left(\frac{\partial}{\partial x_2}\right)^{\alpha_2} \dots \left(\frac{\partial}{\partial x_s}\right)^{\alpha_s}$; $\widetilde{L}(R^k)$ is a set of functions $f(x) = f(x_1, x_2, \dots, x_k)$, such that $\dfrac{f(x)}{(1+|x|^2)^{\frac{k+1}{2}}} \in L(R^k)$.

In the sequel, it will be assumed that the considered functions belong to $\widetilde{L}(R^k)$. We will consider the following generalized mixed derivatives and differentials of arbitrary order of the function $f(x) = f(x_1, x_2, \dots, x_k)$:

1) Let $A \subset M$, $B \subset M$. If there exist functions $a_\beta(\overline{x}_B)$ ($m(\beta) = m(A)$, $|\beta| \le r-1$ if $B = \varnothing$, then $a_\beta(\overline{x}_B) = a_\beta = \text{const}$) and numbers a_α ($m(\alpha) = m(A)$, $|\alpha| = r$), such that there exist limits $\lim\limits_{\overline{x}_B\to\overline{x}_B^0} a_\beta(\overline{x}_B) = a_\beta$ and in the neighborhood of the point x^0 the representation

$$f(x_B + x_{B'}^0 + t_A) = \sum_{i=0}^{r-1} \sum_{|\beta|=i} \frac{a_\beta(\overline{x}_B)}{\beta!} \overline{t}_A^\beta + \sum_{|\alpha|=r} \frac{a_\alpha}{\alpha!} \overline{t}_A^\alpha + \varepsilon(\overline{t}_A, \overline{x}_B) \frac{|\overline{t}_A|^r}{r!}$$

holds; here $\lim\limits_{\substack{|\overline{t}_A|\to 0\\ \overline{x}_B\to\overline{x}_B^0}} \varepsilon(\overline{t}_A, \overline{x}_B) = 0$, and the value $r! \sum\limits_{|\alpha|=r} \frac{a_\alpha}{\alpha!} \overline{t}_A^\alpha$ is called a generalized differential of order r of the function $f(x)$ at the point x^0 with respect to variables whose indices represent the set A, and we denote it by ([65], p. 313; [117], p. 70)

$$d^{(r)}_{A(B)} f(x^0) \Big(d^{(r)}_{A(0)} f(x^0) = d^{(r)}_A f(x^0),$$
$$d^{(r)}_M f(x^0) = d^{(r)} f(x^0); \quad d^{(r)}_{A(M)} f(x^0) = \overline{d}^{(r)}_A f(x^0),$$
$$\overline{d}^{(r)}_M f(x^0) = \overline{d}^{(r)} f(x^0)\Big).$$

The value a_α will be called a generalized mixed derivative of the function $f(x)$ at the point x^0 of order $r = |\alpha|$ with respect to variables whose indices represent the set A, and we denote it by

$$a_\alpha = \mathbf{D}^{(\alpha)}_{A(B)} f(x^0) \Big(\mathbf{D}^{(\alpha)}_{A(\varnothing)} f(x^0) = \mathbf{D}^{(\alpha)}_A f(x^0),$$
$$\mathbf{D}^{(\alpha)}_M f(x^0) = \mathbf{D}^{(\alpha)} f(x^0); \quad \mathbf{D}^{(\alpha)}_{A(M)} f(x^0) = \overline{\mathbf{D}}^{(\alpha)}_A f(x^0),$$
$$\overline{\mathbf{D}}^{(\alpha)}_M f(x^0) = \overline{\mathbf{D}}^{(\alpha)} f(x^0)\Big).$$

2) Let r be an even number. If there exist functions $b_\beta(\overline{x})$ ($|\beta|$ is even, $m(\beta) = m(A)$, $|\beta| \le r-2$ if $B = \varnothing$, then $b_\beta(\overline{x}_\beta) = b_\beta = \text{const}$) and numbers b_α, ($|\alpha| = r$) such that the limits exist $\lim\limits_{\overline{x}_\beta \to \overline{x}_\beta} = b_\beta$, and in the neighborhood of the point x^0 the equality

$$\frac{1}{2}[f(x_{B+x^0_{B'}+t_A}) + f(x_{B+x^0_{B'}-t_A})] = \sum_{i=0}^{\frac{r-2}{2}} \sum_{|\beta|=2i} \frac{1}{\beta!} b_\beta(\overline{x}_\beta) \overline{t}^\beta_A$$
$$+ \sum_{|\alpha|=r} \frac{1}{\alpha!} b_\alpha \overline{t}^\alpha_A + \varepsilon(\overline{t}_A, \overline{x}_B) \frac{|\overline{t}_A|^r}{r!},$$

where $\lim\limits_{\substack{|\overline{t}_A| \to 0 \\ \overline{x}_B \to \overline{x}^0_B}} \varepsilon(\overline{t}_A, \overline{x}_B) = 0$, is valid, then $r! \sum\limits_{|\alpha|=r} \frac{b_\alpha}{\alpha!} \overline{t}^\alpha_A$ is called a generalized symmetric r order differential of the function $f(x)$ at the point x^0 with respect to those variables, whose indices represent the set A, and we denote it by

$$d^{*(r)}_{A(B)} f(x^0) \Big(d^{*(r)}_{A(0)} f(x^0) = d^{*(r)}_A f(x^0);$$
$$d^{*(r)}_M f(x^0) = d^{(r)} f(x^0); \quad d^{*(r)}_{A(M)} f(x^0) = \overline{d}^{*(r)}_A f(x^0);$$
$$\overline{d}^{*(r)}_M f(x^0) = \overline{d}^{*(r)} f(x^0)\Big).$$

When r is odd, we suggest the same definition with the only difference that the sum $f(x_{B+x^0_{B'}+t_A}) + f(x_{B+x^0_{B'}-t_A})$ should be replaced by the difference $f(x_{B+x^0_{B'}+t_A}) - f(x_{B+x^0_{B'}-t_A})$ ([65], p. 315; [117], p. 70).

We call the number b_α a generalized mixed symmetric derivative of the function x^0 at the point $r = |\alpha|$ of order $f(x)$ with respect to variables whose indices represent the set A, and we denote it by

$$b_\alpha = \mathbf{D}^{*(\alpha)}_{A(B)} f(x^0) \ \Big(\mathbf{D}^{*(\alpha)}_{A(\varnothing)} f(x^0) = \mathbf{D}^{*(\alpha)}_{A} f(x^0),$$

$$\mathbf{D}^{*(\alpha)}_{M} f(x^0) = \mathbf{D}^{*(\alpha)} f(x^0); \quad \mathbf{D}^{*(\alpha)}_{A(M)} f(x^0) = \overline{\mathbf{D}}^{*(\alpha)}_{A} f(x^0),$$

$$\overline{\mathbf{D}}^{*(\alpha)}_{M} f(x^0) = \overline{\mathbf{D}}^{*(\alpha)} f(x^0)\Big).$$

The above generalized derivatives are widely used in studying the boundary properties of differentiated Poisson integrals in the space R_+^{k+1}.

4.9 The Boundary Properties of Mixed Derivatives and Differentials of Arbitrary Order of the Poisson Integral for the Half-Space R_+^{k+1} $(k > 1)$

In this section we continue our discussion started in Sections 4.2, 4.4 and 4.7. We prove the Fatou type theorems on the boundary properties of mixed $r \in N$ order derivatives of the Poisson integral for a half-space R_+^{k+1} $(k > 1)$, when the integral density has a mixed derivative or a generalized differential ([108]).

Lemma 4.9.1. *For any $(x, x_{k+1}) \in R_+^{k+1}$, $r \in N$, $A \subset M$, $D \subset M$ and α, $(|\alpha| = r$, $m(\alpha) = m(A))$ the following statements are valid.*

1) $I_\beta^{(\alpha)} = \int\limits_{R^{m(A)}} \dfrac{\partial^r P(t - x, x_{k+1})}{\partial \overline{x}_A^\alpha} \overline{t}_A^\beta d\overline{t}_A = 0$, *where* $m(\beta) = m(\alpha)$, $|\beta| \le |\alpha|$, $\beta \ne \alpha$;

2) $I_r^{(\alpha)} = \dfrac{\Gamma\left(\frac{k+1}{2}\right)}{\pi^{\frac{k+1}{2}}} \int\limits_{R^k} \dfrac{\partial^r P(t - x, x_{k+1})}{\partial \overline{x}_A^\alpha} \dfrac{\overline{t}_A^\alpha}{\alpha!} dt = 1;$

3) $\int\limits_{R^k} \left| \dfrac{\partial^r P(t, x_{k+1})}{\partial \overline{x}_A^\alpha} \right| |t|^r dt < C;$

4) $\int\limits_{R^k} \dfrac{\partial^r P(t - x, x_{k+1})}{\partial \overline{x}_A^\alpha} |\overline{t}_{\mathbf{D}}|^r dt < C$ *for* $\dfrac{x_{k+1}}{|\overline{x}_{\mathbf{D}}|} \ge C > 0;$

5) $\sup\limits_{|t| > \delta > 0} \left| \dfrac{\partial^r P(t, x_{k+1})}{\partial \overline{x}_A^\alpha} \right| \left(|t|^2 + x_{k+1}^2\right)^{\frac{k+1}{2}} |t|^v < C x_{k+1}$, $v = \overline{0, r}$.

Proof. The relation 1) is proved by induction. For $r = 2$, $\alpha_i = \alpha_j = 1$, $i \ne j$, $|\beta| \le 2$, $\beta \ne \alpha$ the equality $I_\beta^{(\alpha)} = 0$ is easy to prove. Assume that for every α $(m(\alpha) = m(A))$ and β $(m(\alpha) = m(\beta)$, $|\beta| \le |\alpha|$, $\beta \ne \alpha)$

$$I_\beta^{(\alpha)} = \int\limits_{R^{m(A)}} \frac{\partial^r P(t - x, x_{k+1})}{\partial \overline{x}_A^\alpha} \overline{t}_A^\beta d\overline{t}_A$$

$$= \int\limits_{R^{m(A)-1}} \overline{t}^{\delta}_{A|v} d\overline{t}_{A|v} \int\limits_{-\infty}^{\infty} \frac{\partial^r P(t-x, x_{k+1})}{\partial \overline{x}^{\alpha}_{A}} t^s_v dt_v = 0,$$

where $s \in \{0, 1, 2, \ldots, |\beta|\}$, $|\delta| = |\beta| - s$ and show that the equality

$$I^{(\gamma)}_{\beta} = \int\limits_{R^{m(A)}} \frac{\partial^{r+1} P(t-x, x_{k+1})}{\partial \overline{x}^{\gamma}_{A}} \overline{t}^{\beta}_{A} d\overline{t}_{A} = 0,$$

where $|\gamma| = |\alpha| + 1$, $|\delta| \leq |\gamma|$ and $\beta \neq \gamma$, holds.

Indeed, using integration by parts, we have

$$\int\limits_{-\infty}^{\infty} \frac{\partial^r P(t-x, x_{k+1})}{\partial \overline{x}^{\alpha}_{A}} t^s_v dt_v = (-1)^q \int\limits_{-\infty}^{\infty} \frac{\partial^r P(t-x, x_{k+1})}{\partial \overline{x}^{\alpha'}_{A|v} \partial t^q_v} t^s_v dt_v$$

$$= \frac{1}{s+1} \int\limits_{-\infty}^{\infty} \frac{\partial^{r+1} P(t-x, x_{k+1})}{\partial \overline{x}^{\alpha'}_{A|v} \partial t^{q+1}_v} t^{s+1}_v dt_v.$$

Therefore we have

$$0 = \frac{1}{s+1} \int\limits_{R^{m(A)-1}} \overline{t}^{\delta}_{A|v} d\overline{t}_{A|v} \int\limits_{-\infty}^{\infty} \frac{\partial^{r+1} P(t-x, x_{k+1})}{\partial \overline{x}^{\alpha'}_{A|v} \partial x^{q+1}_v} t^{s+1}_v dt_v$$

$$= \frac{1}{s+1} \int\limits_{R^{m(A)}} \frac{\partial^{r+1} P(t-x, x_{k+1})}{\partial \overline{x}^{\gamma}_{A}} \overline{t}^{\beta}_{A} d\overline{t}_{A}.$$

Hence $I^{(\gamma)}_{\beta}$, when $|\beta| > 0$, and for $|\beta| = 0$, $I^{(\gamma)}_{\beta} = 0$, as well.

The validity of Statement 2) is also proved by induction. For $r = 2$, $\alpha_i = \alpha_j$, $i \neq j$ the equality $I^{(\alpha)}_2 = 1$ follows from the fact (see Lemma 4.4.2) that

$$I^{(\alpha)}_2 = \frac{\Gamma\left(\frac{k+1}{2}\right)(k+1)(k+3)x_{k+1}}{\pi^{\frac{k+1}{2}}} \int\limits_{R^k} \frac{(t_i - x_i)(t_j - x_j)t_i t_j}{[|t-x|^2 + x^2_{x+1}]^{\frac{k+5}{2}}} dt$$

$$= \frac{\Gamma\left(\frac{k+1}{2}\right)(k+1)(k+3)x_{k+1}}{\pi^{\frac{k+1}{2}}} \int\limits_{R^k} \frac{t^2_i t^2_j dt}{[|t|^2 + x^2_{x+1}]^{\frac{k+5}{2}}} = 1.$$

Assume now that the equality

$$I^{(\alpha)}_n = \frac{\Gamma\left(\frac{k+1}{2}\right)}{\pi^{\frac{k+1}{2}}} \int\limits_{R^k} \frac{\partial^n (t-x, x_{k+1})}{\partial \overline{x}^{\alpha}_{A}} \cdot \frac{\overline{t}^{\alpha}_{A}}{\alpha!} dt = 1$$

is fulfilled for $r = n$ and any $|\alpha| = r$.

Hence we can show that the equality

$$I_{n+1}^{(\gamma)} = \frac{\Gamma\left(\frac{k+1}{2}\right)}{\pi^{\frac{k+1}{2}}} \int\limits_{R^k} \frac{\partial^{n+1} P(t-x, x_{k+1})}{\partial \overline{x}_A^{\gamma}} \cdot \frac{\overline{t}_A^{\gamma}}{\gamma!} dt = 1$$

holds for $|\gamma| = |\alpha| + 1$.

Indeed, using integration by parts, we obtain

$$1 = I_n^{(\alpha)} = \frac{\Gamma\left(\frac{k+1}{2}\right)}{\pi^{\frac{k+1}{2}}} \int\limits_{R^{k-1}} \frac{\overline{t}_{A|v}^{\alpha'}}{\alpha!} d\overline{t}_{A|\nu} \int\limits_{-\infty}^{\infty} \frac{\partial^{n} P(t-x, x_{k+1})}{\partial \overline{x}_A^{\alpha}} t_v^s dt_v$$

$$= \frac{\Gamma\left(\frac{k+1}{2}\right)}{\pi^{\frac{k+1}{2}}} (-1)^{2(s+1)} \int\limits_{R^k} \frac{\partial^{n+1} P(t-x, x_{k+1})}{\partial \overline{x}_A^{\gamma}} \cdot \frac{\overline{t}_A^{\gamma}}{\gamma!} dt = I_{n+1}^{(\gamma)}$$

which was to be shown.

Let us now prove that Statement 3) is valid. We have

$$|t|^r \frac{\partial^r P(t, x_{k+1})}{\partial \overline{t}_A^{\alpha}} = |t|^r x_{k+1} \frac{\partial^r}{\partial \overline{t}_A^{\alpha}} \left[\frac{1}{\left(|t|^2 + x_{k+1}^2\right)^{\frac{k+1}{2}}} \right]$$

$$= |t|^r x_{k+1} \frac{I(t_1, t_2, \ldots, t_k, x_{k+1})}{\left(|t|^2 + x_{k+1}^2\right)^{\frac{k+2r+1}{2}}},$$

where $I(t_1, t_2, \ldots, t_k, x_{k+1})$ is a homogeneous polynomial of degree r in $(t_1, t_2, \ldots, t_k, x_{k+1})$.

Passing to the spherical coordinates, we obtain

$$\int\limits_{R^k} \frac{\partial^r P(t, x_{k+1})}{\partial \overline{t}_A^{\alpha}} |t|^r dt < C x_{k+1} \int\limits_0^{\infty} \frac{T(\rho, x_{k+1}) \rho^{r+k-1}}{\left(\rho^2 + x_{k+1}^2\right)^{\frac{k+2r+1}{2}}} d\rho,$$

where $T(\rho, x_{k+1}) > 0$ is a homogeneous polynomial of degree r in (ρ, x_{k+1}).

Using the substitution $\rho = x_{k+1}\rho_1$, we obtain

$$\int\limits_{R^k} \left| \frac{\partial^r P(t, x_{k+1})}{\partial \overline{t}_A^{\alpha}} \right| |t|^r dt < C \sum_{v=0}^{r} \int\limits_0^{\infty} \frac{\rho_1^{k+r+\nu-1} d\rho_1}{\left(1 + \rho_1^2\right)^{\frac{k+2r+1}{2}}} = O(1).$$

Statement 4) follows from Statement 3) and the fact that

$$|\overline{t}_D + \overline{x}_D|^r \leq 2^r (|\overline{t}_D|^r + |\overline{x}_D|^r).$$

Indeed,

$$\int\limits_{R^k}\Big|\frac{\partial^r P(t-x,x_{k+1})}{\partial\overline{t}^{\alpha}_{A}}\Big|\,|t_{D}|^r dt=\int\limits_{R^k}\Big|\frac{\partial^r P(t,x_{k+1})}{\partial\overline{t}^{\alpha}_{A}}\Big|\,|\overline{t}_{D}+\overline{x}_{D}|^r dt$$

$$\le 2^r\int\limits_{R^k}\left|\frac{\partial^r P(t,x_{k+1})}{\partial\overline{t}^{\alpha}_{A}}\right|\,|t|^r dt+2^r\int\limits_{R^k}\Big|\frac{\partial^r P(t,x_{k+1})}{\partial\overline{t}^{\alpha}_{A}}\Big|\,|\overline{x}_{D}|^r dt$$

$$\le C+C\int\limits_{R^k}\Big|\frac{\partial^r P(t,x_{k+1})}{\partial\overline{t}^{\alpha}_{A}}\Big|\,|\overline{x}_{D}|^r dt.$$

Passing to the spherical coordinates, we obtain

$$I=\int\limits_{R^k}\Big|\frac{\partial^r P(t,x_{k+1})}{\partial\overline{t}^{\alpha}_{A}}\Big|\,|\overline{x}_{D}|^r dt\le Cx_{k+1}|\overline{x}_{D}|^r\int\limits_0^\infty\frac{T(\rho,x_{k+1})\rho^{k-1}d\rho}{(\rho^2+x^2_{k+1})^{\frac{k+2r+1}{2}}},$$

where $T(\rho,x_{k+1})>0$ is a homogeneous polynomial of degree r in (ρ,x_{k+1}).

Using the substitution $\rho=x_{k+1}\rho_1$, we obtain

$$I\le C|\overline{x}_{D}|^r x^{r+k+1}_{k+1}\sum_{v=0}^{r}\int\limits_0^\infty\frac{\rho_1^{k+v-1}d\rho_1}{x^{k+2r+1}_{k+1}(1+\rho_1^2)^{\frac{k+2r+1}{2}}}$$

$$=C\Big(\frac{|\overline{x}_{D}|}{x_{k+1}}\Big)^r\sum_{v=0}^{r}\int\limits_0^\infty\frac{\rho_1^{k+v-1}d\rho_1}{(1+\rho_1^2)^{\frac{k+2r+1}{2}}}<C\quad\text{for}\quad\frac{x_{k+1}}{|x_{D}|}\ge C>0.$$

Thus Statement 4) is proved.

Statement 5) follows from the inequality

$$\Big|\frac{\partial^r P(t,x_{k+1})}{\partial\overline{t}^{\alpha}_{A}}\Big|(|t|^2+x^2_{k+1})^{\frac{k+1}{2}}|t|^v<x_{k+1}\frac{|I(t_1,t_2,\dots,t_k,x_{k+1})|}{(|t|^2+x^2_{k+1})^{r-\frac{v}{2}}},\quad v=\overline{0,r},$$

since $I(t_1,t_2,\dots,t_k,x_{k+1})$ is a homogeneous degree polynomial of degree r in $(t_1,t_2,\dots,t_k,x_{k+1})$.

Lemma 4.9.1 is proved. □

Theorem 4.9.1. *Let $A\subset M$, $B\subset M$ and $B'\subset A$. If $f(x)$ has, at the point x^0, a generalized r-th order differential of $d^{(r)}_{A(B)}f(x^0)$, then for every α ($|\alpha|=r$, $m(\alpha)=m(A)$),*

$$\lim_{(x,x_{k+1})\xrightarrow[\overline{x}_{A|B}]{\wedge}(x^0,0)}\frac{\partial^r U(f;x,x_{k+1})}{\partial\overline{x}^{\alpha}_{A}}=\mathbf{D}^{\alpha}_{A(B)}f(x^0).$$

Proof. Let $x^0=0$. By Statements 1) and 2) of Lemma 4.9.1, we have

$$\frac{\partial^r U(f;x,x_{k+1})}{\partial \overline{x}_A^\alpha}=\frac{\Gamma\left(\frac{k+1}{2}\right)}{\pi^{\frac{k+1}{2}}}\int\limits_{R^k}\frac{\partial^r P(t-x,x_{k+1})}{\partial \overline{x}_A^\alpha}f(t)dt$$

$$=\frac{\Gamma\left(\frac{k+1}{2}\right)}{\pi^{\frac{k+1}{2}}}\int\limits_{R^k}\frac{\partial^r P(t-x+x_{A\cap B},x_{k+1})}{\partial \overline{x}_A^\alpha}f(t+x_{A\cap B})dt$$

$$=\frac{\Gamma\left(\frac{k+1}{2}\right)}{\pi^{\frac{k+1}{2}}}\int\limits_{R^k}\frac{\partial^r P(t-x+x_{A\cap B},x_{k+1})}{\partial \overline{x}_A^\alpha}|\overline{t}_A|^2\Bigg[f(t+x_{A\cap B})$$

$$-\sum_{i=0}^{r-1}\sum_{|\beta|=i}\frac{a_\beta(\overline{x}_{A\cap B},\overline{t}_{A'})}{\beta!}\overline{t}_A^\beta-\sum_{|\beta|=r}\frac{a_\beta}{\beta!}\overline{t}_A^\beta\Bigg]\frac{dt}{|\overline{t}_A|^r}+\mathbf{D}^{(\alpha)}_{A(B)}f(0)$$

$$=C\Bigg(\int\limits_{V_\delta}+\int\limits_{CV_\delta}\Bigg)+\mathbf{D}^{(\alpha)}_{A(B)}f(0)=C(I_1+I_2)+\mathbf{D}^{(\alpha)}_{A(B)}f(0),$$

where V_δ is a ball with center at the point 0 and of radius δ. Let $\varepsilon>0$ and choose $\delta>0$ such that

$$\left|f(t+x_{A\cap B})-\sum_{i=0}^{r-1}\sum_{|\beta|=i}\frac{a_\beta(\overline{x}_{A\cap B},\overline{t}_{A'})}{\beta!}\overline{t}_A^\beta\sum_{|\beta|=r}\frac{\alpha_\beta}{\beta!}\overline{t}_A^\beta\right|\frac{1}{|\overline{t}_A|^r}<\varepsilon$$

for $|t+x_{A\cap B}|<\delta$.

Hence

$$|I_1|\le C\varepsilon\int\limits_{V_\delta}\left|\frac{\partial^r(t-x+x_{A\cap B,x_{k+1}})}{\partial \overline{x}_A^\alpha}\right||\overline{t}_A|^r dt$$

$$\le C\varepsilon\int\limits_{R^k}\left|\frac{\partial^r(t,x_{k+1})}{\partial \overline{x}_A^\alpha}\right|\Bigg(\sqrt{\sum_{i\in A\cap B}t_i^2+\sum_{i\in B'}(t_i+x_i)^2}\Bigg)^r dt$$

$$\le C\varepsilon\int\limits_{R^k}\left|\frac{\partial^r(t,x_{k+1})}{\partial \overline{x}_A^\alpha}\right|\Bigg(\sqrt{\sum_{i\in A\cap B}t_i^2}+\sqrt{\sum_{i\in \beta'}(t_i+x_i)^2}\Bigg)^r dt$$

$$\le C\varepsilon\int\limits_{R^k}\left|\frac{\partial^r(t,x_{k+1})}{\partial \overline{x}_A^\alpha}\right|\Bigg[\Bigg(\sqrt{\sum_{i\in A\cap B}t_i^2}\Bigg)^r+\Bigg(\sqrt{\sum_{i\in \beta'}(t_i+x_i)^2}\Bigg)^r\Bigg]dt$$

$$\le C\varepsilon\Bigg(\int\limits_{R^k}\left|\frac{\partial^r(t,x_{k+1})}{\partial \overline{x}_A^\alpha}\right||t|^r dt+\int\limits_{R^k}\left|\frac{\partial^r P(t,x_{k+1})}{\partial \overline{x}_A^\alpha}\right||(\overline{t+x})_{B'}|^r dt\Bigg),$$

whence by virtue of Statements 3) and 4) of Lemma 4.9.1 it follows that

$$|I_1|<C\varepsilon,\quad\text{for }\frac{x_{k+1}}{|\overline{x}_{B'}|}\ge C>0. \tag{9.1}$$

Furthermore,

$$|I_2| \leq \int\limits_{CV_\delta} \left| \frac{\partial^r P(t - x + x_{A\cap B}, x_{k+1})}{\partial \overline{x}_A^\alpha} \right| \Bigg[|f(t + x_{A\cap B})|$$

$$+ \sum_{i=0}^{r-1} \sum_{|\beta|=i} \frac{|a_\beta(\overline{x}_{A\cap B}, \overline{t}_{A'})|}{\beta!} |t|^i + \Bigg(\sum_{|\beta|=r} \frac{|a_\beta|}{\beta!} \Bigg) |t|^r \Bigg] dt.$$

This inequality, by Statement 5) of Lemma 4.9.1, yields

$$|I_2| < C x_{k+1} \quad \text{for} \quad |x| < \frac{\delta}{2}. \tag{9.2}$$

From (9.1) and (9.2) (assuming ε arbitrary small), it follows that Theorem 4.9.1. is valid. □

Theorem 4.9.2. *Let $A \subset M$, $B \subset M$ and $B' \subset A$. If $f(x)$ at the point x^0 has a generalized symmetric r-th order differential $d^{*(r)}_{A(B)}(x^0)$, then for every α ($|\alpha| = r$, $m(\alpha) = m(A)$)*

$$\lim_{(x_B + x^0_B, x_{k+1}) \to (x^0, 0)} \frac{\partial^r U(f; x_B + x^0_{B'}, x_{k+1})}{\partial \overline{x}_A^\alpha} = \mathbf{D}^{*(\alpha)}_{A(B)} f(x^0).$$

Proof. Let $x^0 = 0$, and r be an even number. We have

$$\frac{\partial^r U(f; x_B, x_{k+1})}{\partial \overline{x}_A^\alpha} = \frac{\Gamma\left(\frac{k+1}{2}\right)}{\pi^{\frac{k+1}{2}}} \int\limits_{R^k} \frac{\partial^r P(t - x + x_{B'}, x_{k+1})}{\partial \overline{x}_A^\alpha} f(t) dt$$

$$= \frac{\Gamma\left(\frac{k+1}{2}\right)}{\pi^{\frac{k+1}{2}}} \int\limits_{R^k} \frac{\partial^r P(t - x + x_A, x_{k+1})}{\partial \overline{x}_A^\alpha} f(t) dt.$$

Note that $\dfrac{\partial^r P(t - x + x_{B'}, x_{k+1})}{\partial \overline{x}_A^\alpha}$ for even r is the even function of variables $\overline{t}_{B'}$, therefore using first the substitution $\overline{t}_{B'} = -\tau_{B'}$ and then $t = \tau + x$, we obtain

$$\frac{\partial^r U(f; x_B, x_{k+1})}{\partial \overline{x}_A^\alpha} = \frac{\Gamma\left(\frac{k+1}{2}\right)}{\pi^{\frac{k+1}{2}}} \int\limits_{R^k} \frac{\partial^r P(t - x + x_{B'}, x_{k+1})}{\partial \overline{x}_A^\alpha} f(t - 2t_{B'}) dt$$

$$= \frac{\Gamma\left(\frac{k+1}{2}\right)}{\pi^{\frac{k+1}{2}}} \int\limits_{R^k} \frac{\partial^r P(t - x + x_A, x_{k+1})}{\partial \overline{x}_A^\alpha} f(t + x_{A\cap B} - 2t_{B'}) dt.$$

The above equalities yield

$$\frac{\partial^r U(f;x_B,x_{k+1})}{\partial \overline{x}_A^\alpha}=\frac{\Gamma\left(\frac{k+1}{2}\right)}{\pi^{\frac{k+1}{2}}}\int\limits_{R^k}\frac{\partial^r P(t-x+x_A,x_{k+1})}{\partial \overline{x}_A^\alpha}$$
$$\times\frac{f(t+x_{A\cap B})+f(t+x_{A\cap B}-2t_{B'})}{2}dt.$$

By Statements 1) and 2) of Lemma 4.9.1, we have

$$\frac{\partial^r U(f;x_B,x_{k+1})}{\partial \overline{x}_A^\alpha}=\frac{\Gamma\left(\frac{k+1}{2}\right)}{\pi^{\frac{k+1}{2}}}\int\limits_{R^k}\frac{\partial^r P(t-x+x_A,x_{k+1})}{\partial \overline{x}_A^\alpha}|\overline{t}_A|^r$$
$$\times\left[\frac{f(t+x_{A\cap B})+f(t+x_{A\cap B}-2t_{B'})}{2}-\sum_{i=0}^{\frac{r-1}{2}}\sum_{|\beta|=i}\frac{b_\beta(\overline{x}_{A\cap B},\overline{t}_{A'})}{\beta!}\overline{t}_A^\beta\right.$$
$$\left.-\sum_{|\beta|=r}\frac{b_\beta}{\beta!}\overline{t}_A^\beta\right]\frac{dt}{|\overline{t}_A|^r}+\mathbf{D}_{A(B)}^{*(\alpha)}f(0).$$

Reasoning as in proving Theorem 4.9.1, from the last equality we conclude that Theorem 4.9.2 is valid. □

Theorem 4.9.3. *If at the point x^0 there exists an r-th order symmetric differential $\overline{d}_A^{*(r)}f(x^0)$, then for any α ($|\alpha|=r$, $m(\alpha)=m(A)$)*

$$\lim_{(x,x_{k+1})\to(x^0,0)}\frac{\partial^r(f;x,x_{k+1})}{\partial\overline{x}_A^{(\alpha)}}=\overline{\mathbf{D}}_A^{*(\alpha)}f(x^0).$$

Proof. Let x^0, and r be an even number. We have

$$\frac{\partial^r U(f;x,x_{k+1})}{\partial \overline{x}_A^\alpha}=\frac{\Gamma\left(\frac{k+1}{2}\right)}{\pi^{\frac{k+1}{2}}}\int\limits_{R^k}\frac{\partial^r P(t-x,x_{k+1})}{\partial \overline{x}_A^\alpha}f(t)dt$$
$$=\frac{\Gamma\left(\frac{k+1}{2}\right)}{\pi^{\frac{k+1}{2}}}\int\limits_{R^k}\frac{\partial^r P(t,x_{k+1})}{\partial \overline{x}_A^\alpha}f(t+x)dt,$$

and also

$$\frac{\partial^r U(f;x,x_{k+1})}{\partial \overline{x}_A^\alpha}=\frac{\Gamma\left(\frac{k+1}{2}\right)}{\pi^{\frac{k+1}{2}}}\int\limits_{R^k}\frac{\partial^r P(t,x_{k+1})}{\partial \overline{x}_A^\alpha}f(x-t)dt.$$

These equalities imply that

$$\frac{\partial^r U(f;x,x_{k+1})}{\partial \overline{x}_A^\alpha} = \frac{\Gamma\left(\frac{k+1}{2}\right)}{\pi^{\frac{k+1}{2}}} \int\limits_{R^k} \frac{\partial^r P(t,x_{k+1})}{\partial \overline{x}_A^\alpha} \frac{f(x+t)+f(x-t)}{2}\, dt.$$

By Statements 1) and 2) of Lemma 4.9.1, we have

$$\frac{\partial^r U(f;x,x_{k+1})}{\partial \overline{x}_A^\alpha} = \frac{\Gamma\left(\frac{k+1}{2}\right)}{\pi^{\frac{k+1}{2}}} \int\limits_{R^k} \frac{\partial^r P(t,x_{k+1})}{\partial \overline{x}_A^\alpha} |\overline{t}_A|^r$$

$$\times \left[\frac{f(x+t)+f(x-t)}{2} - \sum_{i=0}^{\frac{r-2}{2}} \sum_{|\beta|=i} \frac{b_{\beta(x)}}{\beta!} \overline{t}_A^\beta - \sum_{|\beta|=i} \frac{b_\beta}{\beta!} \overline{t}_A^\beta \right] \frac{dt}{|\overline{t}_A|^r} + \overline{\mathbf{D}}_A^{*(\alpha)} f(0).$$

Hence by Statements 3) and 5) of Lemma 4.9.1, we obtain that Theorem 4.9.3 is valid. □

Remark. The examples in Sections 4.2 and 4.4 show that if the type of a generalized derivative changes, then the theorems are not true.

4.10 The Generalized Laplace Operator in R^k $(k \geq 2)$

In this section, we use the notation introduced in Section 3.1.

Let $\Omega(t)$, $t \in R^k$, $|t| = 1$ be a spherical harmonic, i.e. the restriction on the unit sphere of a harmonic polynomial $Q(x) \not\equiv 0$ of degree v, $v = 0,1,\ldots$; $x \in R^k$. Let the function $f(x)$ given in the neighborhood of the point x^0 be integrable on the spheres $|x - x^0| = \rho$ for all sufficiently small $\rho > 0$. If in the neighborhood of the point x^0 we have the equality

$$\frac{1}{|S_\rho^{k-1}|} \int\limits_{S_\rho^{k-1}(x_0)} f(t)\Omega\left(\frac{t}{|t|}\right) dS_\rho^{k-1}(t) = \frac{\Gamma\left(\frac{k}{2}\right)}{2\pi^{\frac{k}{2}}} \int\limits_{s^{k-1}} f(x^0+\rho t)\Omega(t)\, dS^{k-1}(t)$$

$$= \sum_{i=0}^{r} a_i \rho^{v+2i} + o(\rho^{v+2r}) \quad (\rho \to 0+), \tag{10.1}$$

where $dS_\rho^{k-1}(t)$ is an element of the $(k-1)$-dimensional space of the sphere S_ρ^{k-1}, and

$$a_r = \frac{\Gamma\left(\frac{k}{2}\right)}{r!\,2^{v+2r}\,\Gamma\left(\frac{k}{2}+v+r\right)} A_r, \tag{10.2}$$

then the number A_r is called the spherical Laplace Ω-operator of degree r in the function $f(x)$ at the point x^0, and we denote it by $\Omega\overline{\Delta}^r f(x^0)$ (for $k = 2$ and $\Omega(t) = 1$, expansions of the form (10.1) are considered in [67], for $k = 2$ and $\Omega(t) = t_1 + t_2$ in [37], for $k = 2$ and $\Omega(t) = t_1 t_2$ in [38], and in a general case in [39]).

If $\Omega(t) = 1$, the equality (10.1) can be written in the form

$$\frac{1}{|S_\rho^{k-1}|} \int\limits_{S_\rho^{k-1}(x^0)} f(t) dS_\rho^{k-1}(t) = \frac{\Gamma\left(\frac{k}{2}\right)}{2\pi^{\frac{k}{2}}} \int\limits_{S^{k-1}} f(x^0 + \rho t) dS^{k-1}(t)$$

$$= \sum_{i=0}^{r} a_i \rho^{2i} + o(\rho^{2r}) \quad (\rho \to 0).$$

It is not difficult to notice that

$$\overline{\Delta} f(x^0) = \lim_{\rho \to 0} \frac{\frac{1}{|S_\rho^{k-1}|} \int_{S_\rho^{k-1}(x^0)} f(t) dS_\rho^{k-1}(t) - f(x^0)}{\frac{1}{2k}\rho^2}.$$

Let $Q(x)$, $x \in R^k$, $k = 1, 2, \ldots$, be a harmonic homogeneous polynomial of degree v, $v = 0, 1, \ldots$. We say that an integrable function $f(x)$ in the neighborhood of the point $x^0 \in R^k$ has, at this point, a spherical Laplace Q-operator $Q\overline{\Delta}^r f(x^0)$ of order r, $r = 0, 1, 2, \ldots$, if

$$\frac{1}{2\pi^{\frac{k}{2}}} \int\limits_{V^k} f(x^0 + \rho t) Q(t) dt = \sum_{i=0}^{r} b_i \rho^{v+2i} + o(\rho^{v+2r}) \quad (\rho \to 0+),$$

where

$$b_r = \frac{1}{r! 2^{v+2r+1} \Gamma\left(\frac{k}{2} + v + r + 1\right)} Q\overline{\Delta}^r f(x^0).$$

The derivatives $Q\overline{\Delta}^r f(x^0)$ are considered in [27] (p. 495). It is shown there that if an integrable function has a spherical derivative $\Omega\overline{\Delta}^r f(x^0)$ of order r, then it has a globular derivative $Q\overline{\Delta}^r f(x^0)$ of order r with the same value. The converse statement is, generally speaking, invalid, an example of such a function is given.

It is proved in [39] (p. 222), that if a function $f(x)$, $x \in R^k$ has in the neighborhood of the point x^0 all partial derivatives of order $v + 2r + 1$, then

$$\Omega\overline{\Delta}^r f(x_0) = Q(\text{grad}) \delta^r f x_0), \tag{10.3}$$

where the operator $Q(\text{grad})$ is obtained by the substitution of operators $\frac{\partial}{\partial x_i}$, $1 \le i \le k$ in the polynomial $Q(x)$ instead of the coordinates x_i of the point x.

From (10.3) it, in particular, follows ([27], p. 494) that

$$\Omega\overline{\Delta}^r e^{inx} = Q(n) \cdot |n|^{2r} e^{inx} i^{v+2r}.$$

Analogously, we can show that

$$\Omega\overline{\Delta}^r\Big[\frac{1}{(|t|^2+x_{k+1}^2)^{\frac{k+1}{2}}}\Big]=Q(t)\frac{I(t_1,t_2,\dots t_k,x_{k+1})}{(|t|^2+x_{k+1}^2)^{\frac{k+4r+2v+1}{2}}}, \tag{10.4}$$

where $I(t_1,t_2,\dots,t_k,x_{k+1})$ is a homogeneous polynomial of degree $2r$ in $(t_1,t_2,\dots,t_k,x_{k+1})$.

Here we introduce the notion of a spherical Ω and a globular Laplace operator Q in a strong sense.

Let $f(x)$ be a given function in the neighborhood of the point x^0 that is integrable on the spheres $|x-x^0|=\rho$ for all sufficiently small $\rho>0$; B be any subset of the set $M=\{1,2,\dots,k\}$. If there exist functions $a_i(\overline{x}_B)$ (if $B=\varnothing$, then $a_i(\overline{x}_B)=a_i=\text{const}$), $i=\overline{0,r-1}$, given in the neighborhood of x^0 and a number a_r such that there exist limits $\lim\limits_{\overline{x}_B\to\overline{x}_B^0}a_i(\overline{x}_B)=a_i$, and in the neighborhood of x^0,

$$\begin{aligned}&\frac{1}{|S_\rho^{k-1}|}\int\limits_{S_\rho^{k-1}(x_B+x^0_{B'})}f(t)\Omega\Big(\frac{t}{|t|}\Big)dS_\rho^{k-1}(t)\\&=\frac{\Gamma\big(\frac{k}{2}\big)}{2\pi^{\frac{k}{2}}}\int\limits_{S^{k-1}}f(x_B+x^0_{B'}+\rho t)\Omega(t)dS^{k-1}(t)\\&=\sum_{i=0}^{r-1}a_i(\overline{x}_B)\rho^{v+2i}+[a_r+\varepsilon(\rho,\overline{x}_B)]\rho^{v+2r},\end{aligned}$$

where

$$\lim_{\substack{\rho\to0\\ \overline{x}_B\to\overline{x}_B^0}}\varepsilon(\rho,\overline{x}_B)=0,$$

$$a_r=\frac{\Gamma\big(\frac{k}{2}\big)}{r!2^{v+2r}\Gamma\big(\frac{k}{2}+v+r\big)}A_r,$$

then we call the number A_r a spherical Laplace Ω-operator of degree r in the function $f(x)$ at the point x^0 in a strong sense, and we denote it by $\Omega\overline{\Delta}^r_{x_B}f(x^0)$. It is clear that if $B=\varnothing$, then $\Omega\overline{\Delta}^r_{x_B}f(x^0)=\Omega\overline{\Delta}^r f(x^0)$, and if $B=M$, we put $\Omega\overline{\Delta}^r_{x_B}f(x^0)=\Omega\overline{\Delta}^r_x f(x^0)$.

It is obvious that for $\Omega(t)=1$ and $r=1$ we have

$$\overline{\Delta}_{x_B}f(x^0)=\lim_{\substack{\rho\to0\\ x_B+x^0_{B'}\to x^0}}\frac{\frac{1}{|S_\rho^{k-1}|}\int\limits_{S_\rho^{k-1}(x_B+x^0_{B'})}f(t)dS_\rho^{k-1}(t)-f(x_B+x_{B'})}{\frac{1}{2k}\rho^2}.$$

Let there exist functions $b_i(\overline{x}_B)$ given in the neighborhood of the point x^0 (if $B = \varnothing$, then $b_i(\overline{x}_B) = b_i = \text{const}$), $i = \overline{0, r-1}$ and a number b_r such that there exist limits $\lim\limits_{\overline{x}_B \to \overline{x}_B^0} b_i(\overline{x}_B) = b_i$, where B is any subset of the set M.

We say that an integrable function $f(x)$ in the neighborhood of the point $x^0 \in R^k$ has at that point a Laplace globular Q-operator $Q\overline{\Delta}^r_{x_B} f(x^0)$, $r = 0, 1, \dots$ in a strong sense, if

$$\frac{1}{2\pi^{k/2}} \int\limits_{V^k} f(x_B + x^0_{B'} + \rho t) Q(t) dt = \sum_{i=0}^{r-1} b_i(\overline{x}_B)\rho^{v+2i} + [b_r + \varepsilon(\rho, \overline{x}_B)]\rho^{v+2r},$$

where

$$\lim_{\substack{\rho \to 0 \\ \overline{x}_B \to \overline{x}_B^0}} \varepsilon(\rho, \overline{x}_B) = 0,$$

$$b_r = \frac{1}{r! 2^{v+2r+1} \Gamma\left(\frac{k}{2} + v + r + 1\right)} Q\overline{\Delta}^r_{x_B} f(x^0).$$

The notions of Laplace spherical Ω and globular operators Q in R^k are used when dealing with problems of summability of differentiated multiple Fourier series and also when studying the boundary properties of differentiated Poisson integrals in a space R_+^{k+1} $(k > 1)$.

4.11 The Boundary Properties of the Integral $\Omega\Delta^r U(f; x, x_{k+1})$

In this section we prove the Fatou type theorem on the boundary properties of the integral $\Omega\Delta^r U(f; x, x_{k+1})$ when the Poisson integral density has a generalized spherical derivative ([93], [105], [108]–[111]).

The following theorem is easy to prove.

Lemma 4.11.1. *The following statements are valid:*

1) $\int\limits_0^\infty \frac{3\rho^2 - kx_{k+1}^2}{(\rho^2 + x_{k+1}^2)^{\frac{k+5}{2}}} \rho^{k-1} d\rho = 0;$

2) $\frac{(k+1)x_{k+1}\Gamma\left(\frac{k+1}{2}\right)}{k\sqrt{\pi}\Gamma\left(\frac{k}{2}\right)} \int\limits_0^\infty \frac{3\rho^2 - kx_{k+1}^2}{(\rho^2 + x_{k+1}^2)^{\frac{k+5}{2}}} \rho^{k+1} d\rho = 1;$

3) $x_{k+1} \int\limits_0^\infty \frac{|3\rho^2 - kx_{k+1}^2|}{(\rho^2 + x_{k+1}^2)^{\frac{k+5}{2}}} \rho^{k+1} d\rho = O(1).$

Theorem 4.11.1. (a) *Let $B \subset M$. If at the point x^0 there exists a finite derivative $\overline{\Delta}_{x_B} f(x^0)$, then*

$$\lim_{(x_B + x^0_{B'}, x_{k+1}) \to (x^0, 0)} \Delta_x U(f; x_B + x^0_{B'}, x_{k+1}) = \overline{\Delta}_{x_B} f(x^0).$$

(b) *There exists a function $f \in L(R^k)$ for which $\overline{\Delta} f(x^0)$ exists, but the limit*

$$\lim_{(x, x_{k+1}) \xrightarrow{\wedge} (x^0, 0)} \Delta_x U(f; x, x_{k+1})$$

does not exist.

Proof of Item (a). It can be easily verified that

$$\Delta_x U(f; x, x_{k+1}) = -\frac{\partial^2 U(f; x, x_{k+1})}{\partial x^2_{k+1}}$$

$$= \frac{(k+1)x_{k+1}\Gamma\left(\frac{k+1}{2}\right)}{\pi^{\frac{k+1}{2}}} \int_{R^k} \frac{3|t-x|^2 - kx^2_{k+1}}{(|t-x|^2 + x^2_{k+1})^{\frac{k+5}{2}}} f(t)dt.$$

Assume $D_k = \frac{2(k+1)\Gamma\left(\frac{k+1}{2}\right)}{\sqrt{\pi}\Gamma\left(\frac{k}{2}\right)}$ and $\theta = [0, \pi]^{k-2} \times [0, 2\pi]$.

Passing to the spherical coordinates, we obtain

$$\Delta_x U(f; x_B + x_{0_{B'}}, x_{k+1}) = C_k x_{k+1} \int_0^\infty \frac{3\rho^2 - kx^2_{k+1}}{(\rho^2 + x^2_{k+1})^{\frac{k+5}{2}}}$$

$$\times \int_\theta f(x_B + x^0_{B'} + t)\rho^{k-1} \sin^{k-2}\theta_1 \ldots \sin\theta_{k-2} d\rho d\theta_1 \ldots d\theta_{k-2} d\varphi$$

$$= D_k x_{k+1} \int_0^\infty \frac{3\rho^2 - kx^2_{k+1}}{(\rho^2 + x^2_{k+1})^{\frac{k+5}{2}}} \rho^{k-1} \left[\frac{1}{|S^{k-1}_\rho|} \int_{S^{k-1}_\rho (x_B + x^0_{B'})} f(t) dS^{k-1}_\rho(t)\right] d\rho.$$

Hence, by virtue of Lemma 4.11.1, we find that

$$\Delta_x U(f; x_B + x_{0_{B'}}, x_{k+1}) - \overline{\Delta}_{x_B} f(x^0) = D_k x_{k+1} \int_0^\infty \frac{3\rho^2 - kx^2_{k+1}}{(\rho^2 + x^2_{k+1})^{\frac{k+5}{2}}}$$

$$\times \left[\frac{\frac{1}{|S^{k-1}_\rho|} \int_{S^{k-1}_\rho (x_B + x^0_{B'})} f(t) dS^{k-1}_\rho(t) - f(x_B + x^0_{B'})}{\frac{\rho^2}{2k}} - \overline{\Delta}_{x_B} f(x^0)\right] \frac{\rho^2}{2k} d\rho,$$

whence, with Lemma 4.11.1 taken into account, we conclude Item (a) is valid.

The validity of Item (b) follows from Item (b) of Theorem 4.4.2. □

Corollary 4.11.1. *If at the point* x^0 *there exists a finite derivative* $\overline{\Delta}f(x^0)$, *then*

$$\lim_{x_{k+1}\to 0+} \Delta U(f;x^0,x_{k+1}) = \overline{\Delta}f(x^0).$$

Corollary 4.11.2. *If at the point* x^0 *there exists a finite derivative* $\overline{\Delta}_x f(x^0)$, *then*

$$\lim_{(x,x_{k+1})\to(x^0,0)} \Delta x U(f;x^0,x_{k+1}) = \overline{\Delta}_x f(x^0).$$

Using the equality (10.4), analogously to Lemma 4.9.1, we prove the following

Lemma 4.11.2. *For arbitrary integers* $v\ge 0$, $r\ge 1$ *and* $k\ge 1$ *the following statements are valid*

1) $\int\limits_0^\infty \left(\dfrac{\Omega\overline{\Delta}^r_{\rho\theta}P}{\Omega\left(\frac{t}{|t|}\right)}\right)_{x=0} \rho^{v+k+2i-1}d\rho = 0,\ i=\overline{0,r-1};$ *

2) $\dfrac{2\Gamma\left(\dfrac{k+1}{2}\right)}{\sqrt{\pi}r!2^{v+2r}\Gamma\left(\dfrac{k}{2}+v+r\right)} \int\limits_0^\infty \left(\dfrac{\Omega\overline{\Delta}^r_{\rho\theta}P}{\Omega\left(\dfrac{t}{|t|}\right)}\right)_{x=0} \rho^{v+k+2r-1}d\rho = 1;$

3) $\int\limits_0^\infty \left|\left(\dfrac{\Omega\overline{\Delta}^r_{\rho\theta}P}{\Omega\left(\dfrac{t}{|t|}\right)}\right)_{x=0}\right| \rho^{v+k+2r-1}d\rho < C;$

4) $\sup\limits_{\rho\ge\delta>0} \left|\left(\dfrac{\Omega\overline{\Delta}^r_{\rho\theta}}{\Omega\left(\dfrac{t}{|t|}\right)}\right)_{x=0}\right| \rho^{k+1} < Cx_{k+1}$ *for every* $\delta>0$.

5) $\int\limits_0^\infty \left|\left(\dfrac{\Omega\overline{\Delta}^r_{\rho\theta}P}{\Omega\left(\dfrac{t}{|t|}\right)}\right)_{x=0}\right| \rho^{v+k+2i-1}d\rho < Cx_{k+1},\quad i=\overline{0,r}.$

Theorem 4.11.2. *Let* $B\subset M$. *If at the point* x^0 *there exists a finite derivative* $\Omega\overline{\Delta}^r_{x_B} f(x^0)$, *then*

$$\lim_{(x_B+x^0_{B'},x_{k+1})\to(x^0,0)} \Omega\Delta^r U(f;x_B+x^0_{B'},x_{k+1}) = \Omega\overline{\Delta}^r_{x_B} f(x^0).$$

Proof. We have

$$\Omega\Delta^r U(f;x_B+x^0_{B'},x_{k+1}) = \frac{\Gamma\left(\frac{k+1}{2}\right)}{\pi^{\frac{k+1}{2}}} \int\limits_{R^k} \Omega\Delta^r P(t-x_B-x^0_{B'},x_{k+1})f(t)dt$$

$$= \frac{\Gamma\left(\frac{k+1}{2}\right)}{\pi^{\frac{k+1}{2}}} \int\limits_{R^k} \Omega\Delta P|_{x=0}\cdot f(t+x_B+x^0_{B'})dt$$

---

* $\Delta_{\rho\theta}$ is the Laplace operator written in terms of the spherical coordinates $(\rho,\theta_1,\theta_2,\dots,\theta_{k-2},\varphi)$

$$= \frac{\Gamma\left(\frac{k+1}{2}\right)}{\pi^{\frac{k+1}{2}}} \int_0^\infty \overbrace{\int_0^\pi \cdots \int_0^\pi}^{k-2} \int_0^{2\pi} \frac{\Omega\Delta^r_{\rho\theta}P}{\Omega\left(\frac{t}{|t|}\right)}\Big|_{x=0} \Omega\left(\frac{t}{|t|}\right) f(t+x_B+x^0_{B'})\rho^{k-1}$$

$$\times \sin^{k-1}\theta_1 \ldots \sin\theta_{k-2} d\rho d\theta_1 \ldots d\theta_{k-2} d\varphi$$

$$= \frac{\Gamma\left(\frac{k+1}{2}\right)}{\pi^{\frac{k+1}{2}}} \int_0^\infty \frac{\Omega\Delta^r_{\rho\theta}P}{\Omega\left(\frac{t}{|t|}\right)}\Big|_{x=0} |S^{k-1}_\rho| d\rho \frac{1}{|S^{k-1}_\rho|} \int_{S^{k-1}_\rho(x_B+x^0_{B'})} \Omega\left(\frac{t}{|t|}\right) f(t) dS^{k-1}_\rho(t)$$

$$= \frac{2\Gamma\left(\frac{k+1}{2}\right)}{\sqrt{\pi}\Gamma\left(\frac{k}{2}\right)} \int_0^\infty \frac{\Omega\Delta^r_{\rho\theta}P}{\Omega\left(\frac{t}{|t|}\right)}\Big|_{x=0} \rho^{k-1} d\rho \frac{1}{|S^{k-1}_\rho|} \int_{S^{k-1}_\rho(x_B+x^0_{B'})} \Omega\left(\frac{t}{|t|}\right) f(t) dS^{k-1}_\rho(t).$$

By Statements 1) and 2) of Lemma 4.11.2, we obtain

$$\Omega\Delta^r U(f; x_B+x^0_{B'}, x_{k+1}) - \Omega\overline{\Delta}^r_{x_B} f(x^0) = \frac{2\Gamma\left(\frac{k+1}{2}\right)}{\sqrt{\pi}\Gamma\left(\frac{k}{2}\right)} \int_0^\infty \frac{\Omega\Delta^r_{\rho\theta}P}{\Omega\left(\frac{t}{|t|}\right)}\Big|_{x=0} \rho^{k-1}$$

$$\times \left\{ \frac{r!2^{v+2r}\Gamma\left(\frac{k}{2}+v+r\right)}{\Gamma\left(\frac{k}{2}\right)\rho^{v+2r}} \left[\frac{1}{|S^{k-1}_\rho|} \int_{S^{k-1}_\rho(x_B+x^0_{B'})} \Omega\left(\frac{t}{|t|}\right) f(t) dS^{k-1}_\rho(t) \right.\right.$$

$$\left.\left. - \sum_{i=0}^{r-1} a_i(\overline{x}_B)\rho^{v+2i} \right] - \Omega\overline{\Delta}^r_{x_B} f(x^0) \right\} \frac{\Gamma\left(\frac{k}{2}\right)\rho^{v+2r} d\rho}{r!2^{v+2r}\Gamma\left(\frac{k}{2}+v+r\right)}$$

$$= \frac{2\Gamma\left(\frac{k+1}{2}\right)}{\sqrt{\pi} r! 2^{v+2r}\Gamma\left(\frac{k}{2}+v+r\right)} \int_0^\infty \frac{\Omega\Delta^r_{\rho\theta}P}{\Omega\left(\frac{t}{|t|}\right)}\Big|_{x=0}$$

$$\times \left\{ \frac{r!2^{v+2r}\Gamma\left(\frac{k}{2}+v+r\right)}{\Gamma\left(\frac{k}{2}\right)\rho^{v+2r}} \left[\frac{1}{|S^{k-1}_\rho|} \int_{S^{k-1}_\rho(x_B+x^0_{B'})} \Omega\left(\frac{t}{|t|}\right) f(t) dS^{k-1}_\rho(t) \right.\right.$$

$$\left.\left. - \sum_{i=0}^{r-1} a_i(\overline{x}_B)\rho^{v+2i} \right] - \Omega\overline{\Delta}^r_{x_B} f(x^0) \right\} \rho^{v+2r+k-1} d\rho.$$

Hence, by Statements 3), 4) and 5) of Lemma 4.11.2, we conclude that Theorem 4.11.2 is valid. □

Corollary 4.11.3. *If at the point x^0 there exists a finite derivative $\Omega\overline{\Delta}^r f(x^0)$, then*

$$\lim_{x_{k+1}\to 0+} \Omega\Delta^r U(f; x^0, x_{k+1}) = \Omega\overline{\Delta}^r f(x^0).$$

Corollary 4.11.4. *If at the point x^0 there exists a finite derivative $\Omega\overline{\Delta}_x^r f(x^0)$, then*

$$\lim_{(x,x_{k+1})\to(x^0,0)} \Omega\Delta^r U(f;x,x_{k+1}) = \Omega\overline{\Delta}_x^r f(x^0).$$

Remark. An analogue of Theorem 4.11.2 is valid if instead of the Laplace spherical Ω-operator we use a more general globular Laplace Q-operator.

Chapter 5

Boundary Properties of a Differentiated Poisson Integral for a Bicylinder, and Representation of a Function of Two Variables by a Double Trigonometric Series

5.1 Notation and Definitions

Let 2π be a periodic with respect to the variables x and y function $f \in L(Q)$, $(Q = [-\pi, \pi; -\pi, \pi])$ and

$$\sum_{m,n=0}^{\infty} \lambda_{m,n} A_{m,n}(x, y) \tag{1.1}$$

be its Fourier series, where

$$\lambda_{m,n} = \begin{cases} \frac{1}{4}, & \text{when} \quad m = n = 0, \\ \frac{1}{2}, & \text{when} \quad m = 0, \ n > 0 \ \text{or} \ m > 0, \ n = 0, \\ 1, & \text{when} \quad m > 0, \ n > 0, \end{cases}$$

$$A_{m,n}(x, y) = a_{m,n} \cos mx \cos ny + b_{m,n} \sin mx \cos ny$$
$$+ c_{m,n} \cos mx \sin ny + d_{m,n} \sin mx \sin ny,$$

while

$$\left.\begin{aligned}
a_{m,n} &= \frac{1}{\pi^2}\iint\limits_Q f(x,y)\cos mx\cos ny\,dxdy,\\
b_{m,n} &= \frac{1}{\pi^2}\iint\limits_Q f(x,y)\sin mx\cos ny\,dxdy,\\
c_{m,n} &= \frac{1}{\pi^2}\iint\limits_Q f(x,y)\cos mx\sin ny\,dxdy,\\
d_{m,n} &= \frac{1}{\pi^2}\iint\limits_Q f(x,y)\sin mx\sin ny\,dxdy,
\end{aligned}\right\} \tag{1.2}$$

We denote the series (1.1) by $S[f]$.

In a complex form, the Fourier series $S[f]$ can be written more concisely,

$$S[f] = \sum_{m,n=-\infty}^{\infty} c_{m,n} e^{(mx+ny)i}, \tag{1.3}$$

where

$$c_{m,n} = \frac{1}{4\pi^2}\iint\limits_Q f(x,y)e^{-(mx+ny)i}dxdy$$

$$(m = 0, \pm 1, \pm 2, \dots; \quad n = 0, \pm 1, \pm 2, \dots).$$

The abelian means of the series (1.1) are denoted by $U(f;r,\rho,x,y)$ and defined by the equality

$$U(f;r,\rho,x,y) = \sum_{m,n=0}^{\infty} \lambda_{m,n} A_{m,n}(x,y).r^m\rho^n$$

$$(0 < r < 1; \quad 0 < \rho < 1)$$

Taking (1.2) into account, we can easily show that

$$U(f;r,\rho,x,y) = \frac{1}{\pi^2}\iint\limits_Q f(t,\tau)P(r,t-x)P(\rho,\tau-y)dtd\tau$$

$$= \frac{1}{\pi^2}\iint\limits_Q f(x+t,y+\tau)P(r,t)P(\rho,\tau)dtd\tau, \tag{1.4}$$

where

$$P(r,t) = \frac{1}{2} + \sum_{n=1}^{\infty} r^n \cos nt = \frac{1}{2}\cdot\frac{1-r^2}{1-2r\cos t + r^2}.$$

The right-hand side of the equality (1.4) is commonly called the Poisson integral of the function f for a bicylinder. Therefore the expressions “Abelian means of the series $S[f]$” and “the Poisson integral of the function f for a bicylinder” are the synonyms.

5.2 The Boundary Properties of Arbitrary Poisson Integral for a Bicylinder

In this section we consider the question on the validity of Fatou results ([33]) for the Poisson integral in a bicylinder ([87],[88].[90]). We show that in this case no smoothness of density in the neighborhood of some point ensures the existence of boundary values of derivatives in the Poisson integral at the point under consideration. The sufficient conditions for the convergence of derivatives in the Poisson integral are found for a bicylinder. It is proved that the obtained results are unimprovable (in a definite sense).

The following theorem is valid.

Theorem 5.2.1. *No matter what good property the function $f(x,y)$ possesses in the neighborhood of the point (x_0, y_0), the limits*

$$\lim_{(r,\rho)_\lambda \to (1,1)} \frac{\partial U(f;r,\rho,x_0,y_0)}{\partial x}, \qquad \lim_{(r,\rho)_\lambda \to (1,1)} \frac{\partial U(f;r,\rho,x_0,y_0)}{\partial x \partial y}\ {}^{*} \tag{2.1}$$

do not, generally speaking, exist for none of $\lambda > 1$.

Proof. Let $x_0 = y_0 = 0$. Since

$$\frac{\partial P(r, t-x)}{\partial x} = \frac{r(1-r^2)\sin(t-x)}{[1-2r\cos(t-x)+r^2]^2},$$

therefore

$$\frac{\partial U(f;r,\rho,x_0,y_0)}{\partial x} = \frac{r(1-r^2)(1-\rho^2)}{2\pi^2} \iint_Q \frac{\sin(t-x)}{[1-2r\cos(t-x)+r^2]^2} \times \frac{1}{1-2\rho\cos(\tau-y)+\rho^2} f(t,\tau)dtd\tau,$$

whence

$$\frac{\partial U(f;r,\rho,0,0)}{\partial x} = \frac{r(1-r^2)(1-\rho^2)}{2\pi^2} \iint_Q \frac{\sin t}{(1-2r\cos t+r^2)^2} \times \frac{1}{1-2\rho\cos\tau+\rho^2} f(t,\tau)dtd\tau.$$

Let $0 < \delta < \pi$. Assume $D_\delta = [-\pi,\pi;-\pi,\delta] \cup [0,\pi;\delta,\pi]$ and define the function $f(t,\tau)$ as follows:

$$f(t,\tau) = \begin{cases} 0, & \text{when } (t,\tau) \in D_\delta, \\ 1, & \text{when } (t,\tau) \in Q|D_\delta. \end{cases}$$

*The symbol $(r,\rho)_\lambda \to (1,1)$ denotes such tending of r and ρ to 1 under which $1/\lambda \le \frac{1-r}{1-\rho} \le \lambda$, λ is the given number ≥ 1

Clearly, the above-constructed function $f(t,\tau)$ has in the neighborhood of the point $(0,0)$ a derivative of any order, and

$$\frac{\partial U(f;r,\rho,0,0)}{\partial x}=-\frac{r(1-r^2)(1-\rho^2)}{2\pi^2}\int\limits_0^\pi\frac{\sin t dt}{(1-2r\cos t+r^2)^2}\int\limits_\delta^\pi\frac{d\tau}{1-2\rho\cos\tau+\rho^2}$$

$$=-\frac{(1+r)(1+\rho)}{4\pi^2}\int\limits_\delta^\pi\frac{d\tau}{1-2\rho\cos\tau+\rho^2}(1-r)(1-\rho)\Big[\frac{1}{(1-r)^2}-\frac{1}{(1+r)^2}\Big].$$

Hence the limit

$$\lim_{(r,\rho)_\lambda\to(1,1)}\frac{\partial U(f;r,\rho,0,0)}{\partial x}$$

does not exist for $\lambda>1$.

For the constructed function we can also verify that

$$\lim_{(r,\rho)_\lambda\to(1,1)}\frac{\partial U(f;r,\rho,0,0)}{\partial x\partial y}$$

does not exist for $\lambda>1$.

There naturally arises the question: what the sufficient conditions are for the limits (2.1) to exist?

Assume (see Section 4.3)

$$D^2_{xy}f(x,y)=\lim_{(t,\tau)\to(0,0)}\Delta(f;x,y,t,\tau)$$

$$=\lim_{(t,\tau)\to(0,0)}\frac{f(x+t,y+\tau)-f(x,y+\tau)-f(x+t,y)+f(x,y)}{t\tau}.$$

The following lemma is valid. □

Lemma 5.2.1. *Let the derivative $D^2_{x,y}f(x_0,y_0)$ exist and be finite. If $\exists\beta<2$, such that*

$$\operatorname*{Sup}_{\substack{2^i\le\frac{2\pi}{\gamma}\\2^j\le\frac{2\pi}{\delta}}}\frac{1}{\gamma\delta2^{\beta(i+j)}}\int\limits_{-\gamma2^i}^{\gamma2^i}\int\limits_{-\delta2^j}^{\delta2^j}|\Delta(f;x_0,y_0,t,\tau)|dtd\tau=O(1)\tag{2.2}$$

$$(i,j=0,1,2,\dots),$$

then

$$\lim_{(\gamma,\delta)\to(0,0)}A_{\delta\gamma}(f;\sigma,x_0,y_0)=0$$

for $\forall\sigma>\beta$, where

$$A_{\gamma\delta}(f;\sigma,x_0,y_0)=\operatorname*{Sup}_{\substack{2^i\le\frac{2\pi}{\gamma}\\2^j\le\frac{2\pi}{\delta}}}\frac{1}{\gamma\delta2^{\sigma(i+j)}}\int\limits_{-\gamma2^i}^{\gamma2^i}\int\limits_{-\delta2^j}^{\delta2^j}|\Delta(f;x_0,y_0,t,\tau)-D^2_{xy}f(x_0,y_0)|dtd\tau.$$

Proof. Let $\beta < \sigma$. From (2.2), we have

$$\operatorname*{Sup}_{\substack{2^i \le \frac{2\pi}{\gamma} \\ 2^j \le \frac{2\pi}{\delta}}} \frac{1}{\gamma\delta 2^{\beta(i+j)}} \int\limits_{-\gamma 2^i}^{\gamma 2^i} \int\limits_{-\delta 2^j}^{\delta 2^j} |\Delta(f; x_0, y_0, t, \tau) - D^2_{xy} f(x_0, y_0)| dt d\tau = O(1). \tag{2.3}$$

In view of (2.3), for $\forall \varepsilon > 0$ we can choose $N(\varepsilon) > 0$, such that

$$\operatorname*{Sup}_{\substack{2^i \le \frac{2\pi}{\gamma} \\ 2^j \le \frac{2\pi}{\delta}}} \frac{1}{\gamma\delta 2^{\sigma(i+j)}} \int\limits_{-\gamma 2^i}^{\gamma 2^i} \int\limits_{-\delta 2^j}^{\delta 2^j} |\Delta(f; x_0, y_0, t, \tau) - D^2_{xy} f(x_0, y_0)| dt d\tau < \varepsilon, \tag{2.4}$$

is fulfilled, when $i + j \ge N$ for $\forall \gamma$, $\delta > 0$.

Indeed, let $i + j \ge N$. Then

$$\operatorname*{Sup}_{\substack{2^i \le \frac{2\pi}{\gamma} \\ 2^j \le \frac{2\pi}{\delta}}} \frac{1}{\gamma\delta 2^{\sigma(i+j)}} \int\limits_{-\gamma 2^i}^{\gamma 2^i} \int\limits_{-\delta 2^j}^{\delta 2^j} |\Delta(f; x_0, y_0, t, \tau) - D^2_{xy} f(x_0, y_0)| dt d\tau$$

$$= \operatorname*{Sup}_{\substack{2^i \le \frac{2\pi}{\gamma} \\ 2^j \le \frac{2\pi}{\delta}}} \frac{1}{2^{(\sigma-\beta)(i+j)}\gamma\delta 2^{\beta(i+j)}} \int\limits_{-\gamma 2^i}^{\gamma 2^i} \int\limits_{-\delta 2^j}^{\delta 2^j} |\Delta(f; x_0, y_0, t, \tau) - D^2_{xy} f(x_0, y_0)| dt d\tau$$

$$\le \frac{1}{2^{(\sigma-\beta)N}} \operatorname*{Sup}_{\substack{2^i \le \frac{2\pi}{\gamma} \\ 2^j \le \frac{2\pi}{\delta}}} \frac{1}{\gamma\delta 2^{\beta(i+j)}} \int\limits_{-\gamma 2^i}^{\gamma 2^i} \int\limits_{-\delta 2^j}^{\delta 2^j} |\Delta(f; x_0, y_0, t, \tau) - D^2_{xy} f(x_0, y_0)| dt d\tau.$$

Hence it is clear that for the given $\varepsilon > 0$, we can choose $N(\varepsilon) > 0$, such that (2.4) is fulfilled.

On the other hand, when $i + j < N$, we have

$$\operatorname*{Sup}_{i+j<N} \frac{1}{\gamma\delta 2^{\sigma(i+j)}} \int\limits_{-\gamma 2^i}^{\gamma 2^i} \int\limits_{-\delta 2^j}^{\delta 2^j} |\Delta(f; x_0, y_0, t, \tau) - D^2_{xy} f(x_0, y_0)| dt d\tau$$

$$\le \frac{1}{\gamma\delta} \int\limits_{-\gamma 2^N}^{\gamma 2^N} \int\limits_{-\delta 2^N}^{\delta 2^N} |\Delta(f; x_0, y_0, t, \tau) - D^2_{xy} f(x_0, y_0)| dt d\tau. \tag{2.5}$$

Taking into account (2.5) and the fact that

$$\lim_{(t,\tau)\to(0,0)} \Delta(f; x_0, y_0, t, \tau) = D^2_{xy} f(x_0, y_0),$$

for the given $\varepsilon > 0$ we can choose $\eta(\varepsilon) > 0$ such that

$$\operatorname*{Sup}_{i+j<N} \frac{1}{\gamma\delta 2^{\sigma(i+j)}} \int\limits_{-\gamma 2^i}^{\gamma 2^i} \int\limits_{-\delta 2^j}^{\delta 2^j} |\Delta(f; x_0, y_0, t, \tau) - D^2_{xy} f(x_0, y_0)| dt d\tau < \varepsilon, \tag{2.6}$$

when $0 < \gamma,\ \delta < \eta(\varepsilon)$.

The validity of Lemma 5.2.1 follows from (2.4) and (2.6). □

Theorem 5.2.2. *Let $D^2_{xy} f(x_0, y_0)$ exist and be finite. If at the point (x_0, y_0) the condition (2.2) is fulfilled, then*

$$\lim_{\substack{re^{ix} \xrightarrow{\wedge} e^{ix_0} \\ \rho e^{iy} \xrightarrow{\wedge} e^{iy_0}}} \frac{\partial^2 U(f; r, \rho, x, y)}{\partial x \partial y} = D^2_{xy} f(x_0, x_0). \tag{2.7}$$

Proof. It can be easily verified that

$$\frac{1}{r\rho} \cdot \frac{\partial^2 U(f; r, \rho, x, y}{\partial x \partial y}$$

$$= \frac{1}{\pi^2} \iint\limits_Q t\tau \Delta(f; x_0, y_0, t, \tau) K(r, t + x_0 - x) K(\rho, \tau + y_0 - y) dt d\tau,$$

where

$$K(r, u) = \frac{(1 - r^2)\sin u}{(1 - 2r\cos u + r^2)^2}.$$

Without restriction of generality, we may assume that $x_0 = y_0 = 0$.

By the condition

$$\lim_{r \to 1} \frac{1}{\pi} \int\limits_{-\pi}^{\pi} t K(r, t - x) dt = 1, \quad \text{for} \quad \forall x \in (-\pi, \pi),$$

we have

$$\lim_{\substack{(r,x) \xrightarrow{\wedge} (1,0) \\ (\rho,y) \xrightarrow{\wedge} (1,0)}} \frac{\partial^2 U(f; r, \rho, x, y)}{\partial x \partial y} - D^2_{xy} f(0, 0)$$

$$= \frac{1}{\pi^2} \lim_{\substack{(r,x) \xrightarrow{\wedge} (1,0) \\ (\rho,y) \xrightarrow{\wedge} (1,0)}} \iint\limits_Q [\Delta(f; 0, 0, t, \tau) - D^2_{xy} f(0, 0)]$$

$$\times t\tau K(r, t - x) K(\rho, \tau - y) dt d\tau, \tag{2.8}$$

It follows from (2.8) that in order to prove (2.7), it suffices to show that

$$\lim_{\substack{(r,x)\xrightarrow{\wedge}(1,0)\\(\rho,y)\xrightarrow{\wedge}(1,0)}} I(r,\rho,x,y)=\lim_{\substack{(r,x)\xrightarrow{\wedge}(1,0)\\(\rho,y)\xrightarrow{\wedge}(1,0)}}\int_0^\pi\int_0^\pi[\Delta(f;0,0,t,\tau)-D^2_{xy}f(0,0)]$$

$$\times t\tau K(r,t-x)K(\rho,\tau-y)dtd\tau=0. \tag{2.9}$$

We have

$$I(r,\rho,x,y)=\left(\int_0^{1-r}\int_0^{1-\rho}+\int_{1-r}^{\pi}\int_0^{1-\rho}+\int_0^{1-r}\int_{1-\rho}^{\pi}+\int_{1-r}^{\pi}\int_{1-\rho}^{\pi}\right)$$

$$\times[\Delta(f;0,0,t,\tau)-D^2_{xy}f(0,0)]t\tau K(r,t-x)K(\rho,\tau-y)dtd\tau=\sum_{k=1}^{4}I_k(r,\rho,x,y).$$

Using the estimates

$$|K(r,t-x)|\le\frac{2(t+|x|)}{(1-r)^3},\quad 0\le t\le\pi, \tag{2.10}$$

we obtain ([35], p.470)

$$|K(r,t-x)|\le\frac{\pi^3(1-r)}{4r^2(t-|x|)^3},\quad 1-r\le t\le\pi,\quad 2|x|<t, \tag{2.11}$$

$$|I_1(r,\rho,x,y)|<\frac{4[(1-r)(1-\rho)+|x|(1-\rho)+|y|(1-r)+|xy|]}{(1-r)^2(1-\rho)^2}$$

$$\times\int_0^{1-r}\int_0^{1-\rho}|\Delta(f;0,0,t,\tau)-D^2_{xy}f(0,0)|dtd\tau<\frac{C}{4(1-r)(1-\rho)}$$

$$\times\int_{-(1-r)}^{1-r}\int_{-(1-\rho)}^{1-\rho}|\Delta(f;0;0,t,\tau)--D^2_{xy}f(0,0)|dtd\tau<CA_{1-r,1-\rho}(f;\sigma,0,0), \tag{2.12}$$

when

$$\frac{|x|}{1-r}<C,\quad\frac{|y|}{1-\rho}<C.$$

Let μ be an integer satisfying the condition $\pi\le(1-r)2^\mu<2\pi$. Next, not decreasing the generality, we can restrict ourselves to the case

$$\frac{|x|}{1-r}<\frac{1}{2},\quad\frac{|y|}{1-\rho}<\frac{1}{2}.$$

According to those inequalities, with regard for (2.10) and (2.11) ([35], p.469), we have

$$|I_2(r,\rho,x,y)|$$
$$< \frac{\pi^3(1-r)[(1-\rho)+|y|]}{2r^2(1-\rho)^2}\int\limits_{1-r}^{\pi}\int\limits_{0}^{1-\rho}\frac{t}{(t-|x|^3)}|\Delta(f;0,0,t,\tau)-D^2_{xy}f(0,0)|dtd\tau$$
$$< \frac{4\pi^3(1-r)[(1-\rho)+|y|]}{r^2(1-\rho)^2}\int\limits_{1-r}^{\pi}\int\limits_{0}^{1-\rho}\frac{1}{t^2}|\Delta(f;0,0,t,\tau)-D^2_{xy}f(0,0)|dtd\tau$$
$$< \frac{C}{4(1-r)(1-\rho)}\int\limits_{-(1-r)}^{1-r}\int\limits_{-(1-\rho)}^{1-\rho}|\Delta(f;0,0,t,\tau)-D^2_{xy}f(0,0)|dtd\tau$$
$$< \frac{C(1-r)}{1-\rho}\sum_{i=1}^{\mu}\int\limits_{(1-r)2^{i-1}}^{(1-r)2^i}\int\limits_{0}^{\rho}\frac{1}{t^2}|\Delta(f;0,0,t,\tau)-D^2_{xy}f(0,0)|dtdt$$
$$< \frac{C(1-r)}{1-\rho}\sum_{i=1}^{\mu}\frac{4}{(1-r)^2 2^{2i}}\int\limits_{-(1-r)2^i}^{(1-r)2^i}\int\limits_{-(1-\rho)}^{1-\rho}|\Delta(f;0,0,t,\tau)-D^2_{xy}f(0,0)|dtd\tau$$
$$< C\sum_{i=1}^{\mu}\frac{A_{1-r,1-\rho}(f;\sigma,0,0)}{2^{(2-\sigma)}i},$$

whence

$$|I_2(r,\rho,x,y)| < CA_{r,\rho}(f;\sigma,0,0), \tag{2.13}$$

when

$$1<\sigma<2,\quad \frac{|x|}{1-r}<\frac{1}{2},\quad \frac{|y|}{1-\rho}<\frac{1}{2}.$$

Clearly, the same inequality remains for $I_3(r,\rho,x,y)$.

Analogously ([35], p.470),

$$|I_4(r,\rho,x,y)| < \frac{\pi^6(1-r)(1-\rho)}{16r^2\rho^2}\int\limits_{1-r}^{\pi}\int\limits_{1-\rho}^{\pi}\frac{t\tau}{(t-|x|)^3(t-|y|)^3}$$
$$\times|\Delta(f;0,0,t,\tau)-D^2_{xy}f(0,0)|dtd\tau$$
$$< C(1-r)(1-\rho)\sum_{i,j=1}^{\mu}\int\limits_{(1-r)2^{i-1}}^{(1-r)2^j}\int\limits_{(1-\rho)2^{j-1}}^{(1-\rho)2^j}\frac{1}{t^2\tau^2}|\Delta(f;0,0,t,\tau)-D^2_{xy}f(0,0)|dtd\tau$$

$$< C(1-r)(1-\rho)\sum_{i,j=1}^{\mu}\frac{16}{(1-r)^2 2^{2i}(1-\rho)^2 2^{2j}}\int_{-(1-r)2^i}^{(1-r)2^i}\int_{-(1-\rho)2^j}^{(1-\rho)2^j}|\Delta(f;0,0,t,\tau)$$

$$-D^2_{xy}f(0,0)|dtd\tau < C\sum_{i,j=1}^{\mu}\frac{A_{1-r,1-\rho}(f;\sigma,0,0)}{2^{(2-\sigma)i}2^{(2-\sigma)j}}$$

$$< CA_{1-r,1-\rho}(f;\sigma,0,0) \tag{2.14}$$

when

$$1<\sigma<2,\quad \frac{|x|}{1-r}<\frac{1}{2},\quad \frac{|y|}{1-\rho}<\frac{1}{2}.$$

By Lemma 5.2.1, it follows from (2.12),(2.13) and (2.14) that the equality (2.9) is valid, which proves Theorem 5.2.2. □

From Theorem 5.2.2 we obtain a number of corollaries, the most characteristic ones are cited here.

Corollary 5.2.1. *Let $D^2_{xy}f(x_0,y_0)$ exist and be finite. If (2.2) is fulfilled for $\frac{1}{\lambda}\le\frac{\gamma}{\sigma}\le\lambda$, $\lambda\ge1$ then*

$$\lim_{\substack{re^{ix}\xrightarrow{\wedge}e^{ix_0}\\ \rho e^{iy}\xrightarrow{\wedge}e^{iy_0}\\ \frac{1}{\lambda}\le\frac{1-r}{1-\rho}\le\lambda}}\frac{\partial^2U(f;r,\rho,x,y)}{\partial x\partial y}=D^2_{xy}f(x_0,y_0).$$

Corollary 5.2.2. *Let for all $(x,y)\in Q$,*

$$|\Delta(f;x,y,t;\tau)|\le\varphi(x,y),$$

where $\varphi(x,y)$ is finite always everywhere. Then the equality (2.7) is fulfilled almost at all points Q.

Indeed, by Ward's theorem ([116], or [57], p. 212), almost at all points Q there exists $D^2_{xy}f(x,y)$ which is finite. Hence our statement follows from Theorem 5.2.2.

Definition 5.2.1. The point (x,y) is called D-point of the function $f(t,\tau)$, if at that point the conditions

$$\lim_{t,\tau\to0}\frac{1}{t\tau}\int_x^{x+t}\int_y^{y+\tau}f(u,v)dudv=f(x,y),$$

$$\operatorname{Sup}_{|t|>0}\frac{1}{t}\int_x^{x+t}\int_{-\pi}^{\pi}|f(u,v)|dudv=M_1(x)<\infty,$$

$$\operatorname*{Sup}_{|\tau|>0}\frac{1}{\tau}\int_{-\pi}^{\pi}\int_{y}^{y+\tau}|f(u,v)|dudv=M_2(y)<\infty$$

are fulfilled.

As is known, if $f(t,\tau)\in L\ln^+ L$, then almost all points (x,y) of the segment Q are the D-points.

Lemma 5.2.2. *If* (x_0,y_0) *is the D-point of the function* $f(x,y)$*, then at that point the function*

$$F(x,y)=\int_{-\pi}^{x}\int_{-\pi}^{y}f(u,v)dudv$$

satisfies the condition (2.2) *for* $\forall\beta>1$.

Corollary 5.2.3. *If* (x_0,y_0) *is the D-point of the function* $f(x,y)$*, then*

$$\lim_{\substack{re^{ix}\overset{\wedge}{\longrightarrow}e^{ix_0}\\ \rho e^{iy}\overset{\wedge}{\longrightarrow}e^{iy_0}}} U(f;r,\rho,x,y)=f(x_0,y_0).$$

The last statement involves the theorem due to Jessen, Marcinkiewicz and Zygmund ([35] or [36], see also [45]–[47]).

Theorem 5.2.2 is complete in a sense that the following theorem is valid.

Theorem 5.2.3. *Let* $(x_0,y_0)\in Q$ *and* $0<\delta<\min(\pi-x_0,\pi+x_0,\pi-y_0,\pi+y_0)$. *There exists the function* $f(x,y)$ *which is infinitely many times differentiable in the domain* $\Big(-\pi,x_0+\dfrac{\delta}{2};-\pi,y_0+\dfrac{\delta}{2}\Big)$, * *and* $\Delta(f;x_0,t,\tau)\in L(Q)$*, however, the limit*

$$\lim_{(r,\rho)\to(1,1)}\frac{\partial^2U(f;r,\rho,x,y)}{\partial x\partial y}$$

exists at no point $(x,y)\in\{(x_0,y),y\in(-\pi,\pi)\}\cup\{(x,y_0),x\in(-\pi,\pi)\}$.

Proof. Let $x_0=y_0=0$, $0<\delta<\pi$. We define the function $f(t,\tau)$ as follows:

$$f(t,\tau)=\begin{cases}\sqrt{sint}, & \text{when } 0\le t\le\pi,\quad \delta<\tau\le\pi,\\ -\sqrt{-sin\tau}, & \text{when } \delta<t\le\pi,\quad -\pi\le\tau\le0,\\ 0, & \text{when } (t,\tau)\in Q|\{(0,\pi;\delta,\pi)\cup(\delta<t\le\pi;-\pi\le\tau\le0)\}.\end{cases}$$

*We consider the case $0\le x_0<\pi$ and $0\le y_0<\pi$.

For that function $\Delta(f;0,0,t,\tau)\in L(Q)$, in the domain $(-\pi,\delta;-\pi,\delta)$ this function has derivatives of any order, and for every point $(x,y)\in(-\pi,\delta;-\pi,\delta)$, $d^k f(x,y) = 0$, $k\geq 1$ *. For every point $(0,y)$, $-\pi<y<\pi$ we have

$$\frac{\partial^2 U(f;t,\rho,0,y)}{\partial x\partial y}=\frac{r\rho(1-r^2)(1-\rho^2)}{\pi^2}\left\{\int_0^\pi\int_\delta^\pi\frac{\sin t}{(1-2r\cos t+r^2)^2}\right.$$

$$\times\frac{\sin(\tau-y)}{[1-2\rho\cos(\tau-y)+\rho^2]^2}\sqrt{\sin t}dtd\tau-\int_\delta^\pi\int_{-\pi}^0\frac{\sin t}{(1-2r\cos t+r^2)^2}$$

$$\left.\times\frac{\sin(\tau-y)}{[1-2\rho\cos(\tau-y)+\rho^2]^2}\sqrt{-\sin\tau}dtd\tau\right\}$$

$$=\frac{r\rho(1-r^2)(1-\rho^2)}{\pi^2}\left\{\int_0^\pi\frac{\sin^{\frac{3}{2}}tdt}{(1-2r\cos t+r^2)^2}\int_\delta^\pi\frac{\sin(\tau-y)d\tau}{[1-2\rho\cos(\tau-y)+\rho^2]^2}\right.$$

$$\left.+\int_\delta^\pi\frac{\sin tdt}{(1-2r\cos t+r^2)^2}\int_0^\pi\frac{\sqrt{\sin\tau}\sin(\tau+y)d\tau}{[1-2\rho\cos(\tau+y)+\rho^2]^2}\right\}.$$

Consider the integral

$$I=\int_0^\pi\frac{\sin^{\frac{3}{2}}xdx}{(1-2r\cos x+r^2)^2}=\int_0^\pi\frac{\sin^{\frac{3}{2}}xdx}{\left[(1-r)^2+4r\sin^2\frac{x}{2}\right]^2}.$$

By the substitution $t=\operatorname{tg}\frac{x}{2}$, we obtain

$$I=2^{\frac{5}{2}}\int_0^\infty\frac{t^{\frac{3}{2}}dt}{(1+t^2)^{\frac{1}{2}}[(1-r)^2+(1+r)^2t^2]^2}$$

$$=\frac{2^{\frac{3}{2}}}{(1-r)^{\frac{3}{2}}}\int_0^\infty\frac{t^{\frac{1}{4}}dt}{\sqrt{1+(1-r)^2t}[1+(1+r)^2t]^2}.$$

Hence it is not difficult to see that the limit

$$\lim_{(r,\rho)\to(1,1)}\frac{\partial^2 U(r,\rho,0,y)}{\partial x\partial y}$$

does not exist.

Analogously, we can show that for every point $(x,0)$, $-\pi<x<\pi$ the limit

$$\lim_{(r,\rho)\to(1,1)}\frac{\partial^2 U(f;r,\rho,x,0)}{\partial x\partial y}$$

*Where $d^k(f(x,y)$ is the total k-th order differential of the function $f(x,y)$

does not exist.

Assume now (see Section 4.3)

$$g(f;x,y,t,\tau)$$
$$\frac{f(x+t,y+\tau)-f(x+t,y-\tau)-f(x-t,y+\tau)+f(x-t,y-\tau)}{4t\tau}.$$

If there exists the limit $\lim_{t,\tau\to 0} g(f;x,y,t,\tau)$, then this limit is called a symmetric derivative of the function $f(x,y)$ at the point (x,y), and we denote it by $D^*_{xy}f(x,y)$. □

Reasoning as above, we can prove the following theorems.

Theorem 5.2.4. (a) *Let $D^*_{xy}f(x_0,y_0)$ exist and be finite. If $\exists\beta<2$, such that*

$$\operatorname*{Sup}_{\substack{2^i\le\frac{2\pi}{\gamma}\\ 2^j\le\frac{2\pi}{\delta}}} \frac{1}{\gamma\delta 2^{\beta(i+j)}}\int\limits_{-\gamma 2^i}^{\gamma 2^i}\int\limits_{-\delta 2^j}^{\delta 2^j}|g(f;x_0,y_0,t,\tau)|dtd\tau=O(1), \tag{2.15}$$

then

$$\lim_{(r,\rho)\to(1,1)}\frac{\partial^2 U(f;r,\rho,x_0,y_0)}{\partial x\partial y}=D^*_{xy}f(x_0,y_0).$$

(b) *There exists the function $f(t,\tau)$, such that $D^*_{xy}f(x_0,y_0)=0$, and the condition* (2.15) *is fulfilled, however, there is no limit*

$$\lim_{\substack{re^{ix}\xrightarrow{\wedge}e^{ix_0}\\ \rho e^{iy}\xrightarrow{\wedge}e^{iy_0}}}\frac{\partial^2 U(f;r,\rho,x,y)}{\partial x\partial y}.$$

Theorem 5.2.5. *Let*

$$\lim_{(x,y)\to(x_0,y_0)}\frac{f(x,y)-f(x_0,y)}{x-x_0}=D_xf(x_0,y_0).$$

exist and be finite. If $\exists\beta<2$, such that

$$\operatorname*{Sup}_{\substack{2^i\le\frac{2\pi}{\gamma}\\ 2^j\le\frac{2\pi}{\delta}}} \frac{1}{\gamma\delta 2^{\beta(i+j)}}\int\limits_{-\gamma 2^i}^{\gamma 2^i}\int\limits_{-\delta 2^j}^{\delta 2^j}\left|\frac{f(x_0+t,y)-f(x_0,y)}{t}\right|dtdy=O(1), \tag{2.16}$$

then

$$\lim_{\substack{re^{ix}\xrightarrow{\wedge}e^{ix_0}\\ \rho e^{iy}\xrightarrow{\wedge}e^{iy_0}}}\frac{\partial U(f;r,\rho,x,y)}{\partial x}=D_xf(x_0y_0).$$

Theorem 5.2.6. (a) *Let*

$$\lim_{x\to x_0}\frac{f(x,y)-f(x_0,y)}{x-x_0}=\frac{\partial f(x_0,y)}{\partial x}$$

uniformly with respect to $\frac{\partial f(x_0,y)}{\partial x}\in L(-\pi,\pi)$ *in some neighborhood of the point* y_0, *and* y_0 *is the Lebesgue point of the function* $\frac{\partial f(x_0,y)}{\partial x}$. *If, moreover, the condition* (2.16) *is fulfilled, then*

$$\lim_{\substack{re^{ix}\xrightarrow{\wedge}e^{ix_0}\\ \rho e^{iy}\xrightarrow{\wedge}e^{iy_0}}}\frac{\partial U(f;r,\rho,x,y)}{\partial x}=\frac{\partial f(x_0,y_0)}{\partial x}.$$

(b) *Let* $(x_0,y_0)\in Q$ *and* $0<\delta<\min(\pi-y_0,\pi+y_0)$. *There exists the function* $f(x,y)$ *which is infinitely many times differentiable in the domain* $\left(-\pi,\pi;-\pi,y_0+\frac{\delta}{2}\right)^*$ *and* $\frac{f(x,y)-f(x_0,y)}{x-x_0}\in L(Q)$, *however, the limit*

$$\lim_{(r,\rho)\to(1,1)}\frac{\partial U(f;r,\rho,x_0,y)}{\partial x}$$

exists at no point (x_0,y), $-\pi<y<\pi$.

Remark. Theorem 5.2.2 and Item (a) of Theorem 5.2.4 are valid, respectively, for the derivatives

$$C_1D^2_{xy}f(x,y)=\lim_{\substack{h\to 0\\ l\to 0}}\frac{4}{h^2l^2}\int_0^h\int_0^l[f(x+t,y+\tau)-f(x+t,y)$$
$$-f(x,y+\tau)+f(x,y)]dtd\tau,$$

$$C_1D^*_{xy}f(x,y)=\lim_{\substack{h\to 0\\ l\to 0}}\frac{1}{h^2l^2}\int_0^h\int_0^l[f(x+t,y+\tau)-f(x-t,y+\tau)$$
$$-f(x+t,y-\tau)+f(x-t,y-\tau)]dtd\tau.$$

*We consider the case $0\le y_0<\pi$.

5.3 Representation of a Function of Two Variables by a Trigonometric Series in the Case of Spherical Convergence

This section is devoted to the consideration of the problem of representability of a measurable and always everywhere finite function of two variables by a double trigonometric series in the case of spherical convergence; namely, we prove the theorem, analogous to that proven by N.N. Luzin ([43], p.236) for the function of one variable.

The spherical δ order Riesz means of the series (1.3) are called the sum

$$S_R^\delta(f;x,y)=\sum_{\sqrt{m^2+n^2}\le R}\Big(1-\frac{m^2+n^2}{R^2}\Big)^\delta c_{m,n}e^{(mx+ny)i}\quad(\delta\ge 0,\ R>0).$$

If $\lim\limits_{R\to\infty}S_R^\delta(f;x,y)=S$, then they say that the series (1.3) at the point (x,y) is summable to the number S by the Riesz method with exponent δ.

Assume

$$A_\varepsilon(f;x,y)=\sum_{m,n=0}^{\infty}c_{m,n}e^{[(mx+ny)i-\varepsilon\sqrt{m^2+n^2}]}.$$

If there exists a finite limit

$$\lim_{\varepsilon\to0}A_\varepsilon(f;x,y)=S,$$

then they say that the series (1.3) at the point (x,y) is summable to the number S by the Abel–Poisson method.

From the theorem proven in [117], as a consequence we obtain the following

Theorem A. *Let $f(x,y)\in L(Q)$ $(Q=[-\pi,\pi,-\pi,\pi])$ be of period 2π with respect to every argument, and at the point (x,y) have a total differential, then*

1) $\lim\limits_{\varepsilon\to0}\dfrac{\partial}{\partial x}A_\varepsilon(f;x,y)=\dfrac{\partial f}{\partial x}$ *and* $\lim\limits_{\varepsilon\to0}\dfrac{\partial}{\partial y}A_\varepsilon(f;x,y)=\dfrac{\partial f}{\partial y}$,

2) $\lim\limits_{R\to\infty}\dfrac{\partial}{\partial x}S_R^\delta(f;x,y)=\dfrac{\partial f}{\partial x}$ *and* $\lim\limits_{R\to\infty}\dfrac{\partial}{\partial y}S_R^\delta(f;x,y)=\dfrac{\partial f}{\partial y}$ *for* $\delta>\dfrac{3}{2}$.

The following theorem is valid ([94]).

Theorem 5.3.1. *Let $f_1(x,y)$ and $f_2(x,y)$ be arbitrary measurable and almost everywhere finite functions on Q. Then there exists the continuous function $F(x,y)$, such that if* (1.3) *is its Fourier series, then almost everywhere on Q we have*

1) $\lim\limits_{\varepsilon\to0}\dfrac{\partial}{\partial x}A_\varepsilon(F;x,y)=f_1(x,y)$ *and* $\lim\limits_{\varepsilon\to0}\dfrac{\partial}{\partial y}A_\varepsilon(F;x,y)=f_2(x,y)$,

2) $\lim\limits_{R\to\infty}\dfrac{\partial}{\partial x}S_R^\delta(F;x,y)=f_1(x,y)$ *and* $\lim\limits_{R\to\infty}\dfrac{\partial}{\partial y}S_R^\delta(F;x,y)=f_2(x,y)$ *for* $\delta>\dfrac{3}{2}$.

Proof. Let $f_1(x,y)$ and $f_2(x,y)$ be arbitrary measurable and almost everywhere finite functions on Q. By A. Jvarshishvili's theorem ([14]), there exists the continuous function $F(x,y)$, such that almost everywhere on Q,

$$dF(x,y) = f_1(x,y)dx + f_2(x,y)dy.$$

Let (1.3) be the Fourier series of the function $F(x,y)$, then by Theorem A, almost everywhere on Q:

1) $\lim\limits_{\varepsilon\to 0} \dfrac{\partial}{\partial x} A_\varepsilon(F;x,y) = f_1(x,y)$ and $\lim\limits_{\varepsilon\to 0} \dfrac{\partial}{\partial y} A_\varepsilon(F;x,y) = f_2(x,y)$,

2) $\lim\limits_{R\to\infty} \dfrac{\partial}{\partial x} S_R^\delta(F;x,y) = f_1(x,y)$ and $\lim\limits_{R\to\infty} \dfrac{\partial}{\partial y} S_R^\delta(F;x,y) = f_2(x,y)$ for $\delta > \dfrac{3}{2}$.

Thus Theorem 5.3.1 is complete. □

5.4 On One Method of Summation of Double Fourier Series

In this section we consider the method of summation of double Fourier series. The method allows one to establish not only summability almost everywhere, but also to show a set of points of total measure at which the summability takes place.

Let the function $f(x,y)$ be summable on $Q = [-\pi,\pi;-\pi,\pi]$, and 2π-periodic with respect to every variable. Assume (see Section 4.10)

$$\Delta_r(f;P) = \frac{\frac{1}{2\pi r}\int_{C(P;r)} f(t,\tau)ds(t,\tau) - f(x,y)}{\frac{1}{4}r^2},$$

where $C(P;r)$ is the circumference of radius r, with center at the point $P(x,y)$. The generalized Laplace operator ([74], p.61; [55], p.279) $\overline{\Delta}f(P)$ of the function $f(P) = f(x,y)$ at the point $P(x,y)$ is defined by the equality

$$\overline{\Delta}f(P) = \lim_{r\to 0} \Delta_r(f;P).$$

The operator Ω on Q is defined ([55], p.281); [56],p.293) by the equality

$$\Omega f(x,y) = \Omega_Q f(x,y) = -\frac{1}{2\pi}\iint\limits_Q f(t,\tau)g(x,y;t,\tau)dtd\tau,$$

where $g(x,y;t,\tau)$ is the Green's function on Q.

The point $P(x,y)$ is called the L-point of the function $f(t,\tau)$, if

$$\iint\limits_{I(P;r)} |f(t,\tau) - f(x,y)|dtd\tau = o(r^2) \quad \text{for} \quad r\to 0,$$

where $I(P;r)$ is a circle of radius r, with center at the point $P(x,y)$.

As is known, if $f \in L(Q)$, then almost all points (x,y) of the segment Q are the L-points of the function $f(t,\tau)$.

The following lemma is valid ([55], p. 282; [56], p. 296).

Lemma A. *Let $f \in L(Q)$ and $P(x,y)$ be the L-points of the function $f(t,\tau)$. Then $\Omega f(x,y)$ is finite almost everywhere on Q, and*

$$\overline{\Delta}\Omega f(x,y) = f(x,y).$$

Further,

$$\begin{aligned}\overline{\Delta}(\sin mx \cos ny) &= \frac{\partial^2}{\partial x^2}(\sin mx \cos ny) + \frac{\partial^2}{\partial y^2}(\sin mx \cos ny)\\ &= -(m^2+n^2)\sin mx \cos ny,\end{aligned}$$

whence ([56], p.294)

$$\Omega\overline{\Delta}(\sin mx \cos ny) = -(m^2+n^2)\Omega(\sin mx \cos ny) = \sin mx \cos ny.$$

Consequently,

$$\Omega(\sin mx \cos ny) = -\frac{1}{m^2+n^2}\sin mx \cos ny.$$

Consider the double trigonometric series (1.1). Of the terms of the series (1.1) we compose the series

$$\begin{aligned}-\frac{1}{2}\sum_{m=1}^{\infty}\frac{a_{m,0}\cos mx + b_{m,0}\sin mx}{m^2} - \frac{1}{2}\sum_{n=1}^{\infty}\frac{a_{0,n}\cos ny + c_{0,n}\sin ny}{n^2}\\ - \sum_{m,n=1}\frac{A_{m,n}(x,y)}{m^2+n^2}.\end{aligned} \tag{4.1}$$

Assume that (4.1) is the Fourier series of the function $F \in L(Q)$.

Definition 5.4.1. We call the series (1.1) R^* summable to $\frac{1}{4}a_{0,0} + S(x,y)$ at the point $P(x,y)$, if

$$\overline{\Delta}F(x,y) = S(x,y).$$

The following lemma is valid.

Lemma 5.4.1. *Let $f \in L(Q)$ and*

$$S[f] = \frac{1}{2}\sum_{m=1}^{\infty}(a_{m,0}\cos mx + b_{m,0}\sin mx)$$

$$+\frac{1}{2}\sum_{n=1}^{\infty}(a_{0,n}\cos ny + c_{0,n}\sin ny) + \sum_{m,n}^{\infty} A_{m,n}(x,y), \tag{4.2}$$

then $\Omega f \in L(Q)$, and the series (4.1) is its Fourier series.

Proof. Indeed, as is known ([56], p.294), $\Omega f \in L(Q)$, and hence

$$\iint_Q \Omega f(x,y)\sin mx\cos ny dxdy$$

$$= -\frac{1}{2\pi}\iint_Q \sin mx\cos ny dxdy \iint_Q f(t,\tau)g(x,y;t,\tau)dtd\tau$$

$$= -\frac{1}{2\pi}\iint_Q f(t,\tau)dtd\tau \iint_Q \sin mx\cos ny g(x,y;t,\tau)dtd\tau$$

$$= \iint_Q f(t,\tau)\Omega(\sin mt\cos n\tau)dtd\tau$$

$$= -\frac{1}{m^2+n^2}\iint_Q f(t,\tau)\sin mt\cos n\tau dtd\tau.$$

Consequently,

$$b_{m,n}(\Omega, f) = -\frac{b_{m,n}(f)}{m^2+n^2}.$$

Equalities for another coefficients are proved analogously. Thus the lemma is proved. □

Theorem 5.4.1. *For every summable function $f(x,y)$, the series $S[f]$ is R^*-summable at all L-points of the function to $f(x,y)$.*

Proof. Let (4.2) be the Fourier series of the function f. Then according to Lemma 5.4.1, the series (4.1) is the Fourier series of the function Ωf.

Assume now that the point (x,y) is the L-point of the function $f(t,\tau)$. Then by Lemma A,

$$\overline{\Delta}\Omega f(x,y) = f(x,y).$$

Thus the theorem is complete. □

5.5 Representation of a Function of Two Variables by a Double Trigonometric Series in the Case of Ptingsheim's Convergence

In this section we consider the problem of representability of a function of two variables by a double trigonometric series, and also in the case of Pringsheim's convergence.

Regarding this problem, the function of many variables is, as L. Zhizhiashvili says in his review paper ([29], p. 76), very little studied.

The question on the representation of measurable functions of two variables by a double trigonometric series has been considered by A. Jvarsheishvili ([11]. O. Dzagnidze's work [16] deals with the same problem.

In this section, we establish the existence of a continuous function $F(x, y)$ of two variables, whose differentiated Fourier series is, depending on a way of differentiation, summable almost at every point to different arbitrarily fixed and independent of each other measurable functions.

We will use the following derivatives of the function of two variables $f(x, y)$:

$$D_{x(y)}f(x_0, y_0) = \lim_{(t,y)\to(0,y_0)} \frac{f(x_0 + t, y) - f(x_0, y)}{t},$$

$$D_{(x)y}f(x_0, y_0) = \lim_{(x,\tau)\to(x_0,0)} \frac{f(x, y_0 + \tau) - f(x, y_0)}{\tau},$$

$$D^*_{x(y)}f(x_0, y_0) = \lim_{(t,y)\to(0,y_0)} \frac{f(x_0 + t, y) - f(x_0 - t, y)}{2t},$$

$$D^*_{(x)y}f(x_0, y_0) = \lim_{(x,\tau)\to(x_0,0)} \frac{f(x, y_0 + \tau) - f(x, y_0 - \tau)}{2\tau}$$

(see Section 4.1),

$$D^2_{xy}f(x_0, y_0)$$
$$= \lim_{(t,\tau)\to(0,0)} \frac{f(x_0 + t, y_0 + \tau) - f(x_0 + t, y_0) - f(x_0, y_0 + \tau) + f(x_0, y_0)}{t\tau}$$

(see Section 4.3),

$$\overline{\Delta} f(x_0, y_0) = \lim_{r\to 0} \frac{\frac{1}{2\pi r}\int_{C(x_0,y_0;r)} f(t,\tau)ds(t,\tau) - f(x_0, y_0)}{\frac{1}{4}r^2}$$

(see Section 4.10).

Assume

$$\Delta_{22}(f; x, y, h, l) = \frac{1}{h^2 l^2}\int_0^h\int_0^l [f(x + t, y + \tau) - f(x + t, y - \tau)$$

$$-f(x-t,y+\tau)+f(x-t,y-\tau)]dtd\tau,$$

$$\Delta_{21}(f;x,y,h,l)=\frac{1}{2h^2l}\int_0^h\int_0^l[f(x+t,y+\tau)-f(x-t,y+\tau)$$
$$+f(x+t,y-\tau)-f(x-t,y-\tau)]dtd\tau,$$

$$\Delta_{12}(f;x,y,h,l)=\frac{1}{2hl^2}\int_0^h\int_0^l[f(x+t,y+\tau)-f(x+t,y-\tau)$$
$$+f(x-t,y+\tau)-f(x-t,y-\tau)]dtd\tau.$$

Let

$$D_{22}f(x,y)=\lim_{(h,l)\to(0,0)}\Delta_{22}(f;x,y,h,l),$$
$$D_{21}f(x,y)=\lim_{(h,l)\to(0,0)}\Delta_{21}(f;x,y,h,l),$$
$$D_{12}f(x,y)=\lim_{(h,l)\to(0,0)}\Delta_{12}(f;x,y,h,l),$$

We can easily verify that the following lemma is valid.

Lemma 5.5.1. 1) *From the existence of* $D_{x(y)}f(x_0,y_0)$ $(D_{(x)y}f(x_0,y_0))$ *follows that of* $D^*_{x(y)}f(x_0,y_0)$ $(D^*_{(x)y}f(x_0,y_0))$, *and* $D^*_{x(y)}f(x_0,y_0)=D_{x(y)}f(x_0,y_0)$ $(D^*_{(x)y}f(x_0,y_0)=D_{(x)y}f(x_0,y_0))$.

2) *From the existence of* $D^*_{x(y)}f(x_0,y_0)$ $(D^*_{(x)y}f(x_0,y_0))$ *follows that of* $D_{21}f(x_0,y_0)$ $(D_{12}f(x_0,y_0))$, *and* $D_{21}f(x_0,y_0)=D^*_{x(y)}f(x_0,y_0)$ $(D_{12}f(x_0,y_0)=D^*_{(x)y}f(x_0,y_0))$.

3) *From the existence of* $D^2_{xy}f(x_0,y_0)$ *follows that of*

$$D_{22}f(x_0,y_0)\quad\textit{and}\quad D_{22}f(x_0,y_0)=D^2_{xy}f(x_0,y_0).$$

Analogously to Theorem 3.6.3, we can prove the following ([84])

Theorem 5.5.1. *Let* $f_i(x,y)$, $i=1,2,3,4$, *be arbitrary measurable and almost everywhere finite function on* $Q=[0,2\pi;0,2\pi]$. *Then there exists the continuous function* $F(x,y)$, *such that almost everywhere on* Q,

$$D_{x(y)}F(x,y)=f_1(x,y);\quad D_{(x)y}F(x,y)=f_2(x,y);$$
$$D^2_{xy}F(x,y)=f_3(x,y);\quad \overline{\Delta}F(x,y)=f_4(x,y);$$

Assume there is a double trigonometric series

$$\sum_{m,n=0}^{\infty}\lambda_{m,n}A_{m,n}(x,y),\tag{5.1}$$

where

$$\lambda_{m,n}=\begin{cases}\frac{1}{4} & \text{for } \ m=n=0,\\ \frac{1}{2} & \text{for } \ m=0, n>0, \quad \text{or} \quad m>0, \quad n=0,\\ 1 & \text{for } \ m\geq 1, \quad n\geq 1,\end{cases}$$

$$A_{m,n}(x,y)=a_{m,n}\cos mx\cos ny+b_{m,n}\sin mx\cos ny$$
$$+c_{m,n}\cos mx\sin ny+d_{m,n}\sin mx\sin ny,$$

The series (5.1) is called $R(\alpha\beta)$-summable (α and β are natural numbers) to $S(x,y)$ at the point (x,y), if

$$\lim_{(u,v)\to(0,0)} R_{\alpha,\beta}(x,y,u,v)=S(x,y),$$

where

$$R_{\alpha,\beta}(x,y,u,v)=\sum_{m,n=0}^{\infty}\lambda_{m,n}A_{m,n}(x,y)\Big(\frac{\sin mu}{mu}\Big)^{\alpha}\Big(\frac{\sin nv}{nv}\Big)^{\beta}.$$

We will consider the methods $R(2,2)$, $R(2,1)$, $R(1,2)$ and R^*. The method $R(2,2)$ has been considered in [7], [11], [12], [28], [72] and [73].

Of the terms of the series (5.1) we compose new double series

$$\frac{a_{0,0}}{16}x^2y^2-\frac{y^2}{4}\sum_{m=1}^{\infty}\frac{A_{m,0}(x,y)}{m^2}-\frac{x^2}{4}\sum_{n=1}^{\infty}\frac{A_{0,n}(x,y)}{n^2}+\sum_{m,n=1}^{\infty}\frac{A_{m,n}(x,y)}{m^2n^2}; \tag{5.2}$$

$$\frac{a_{0,0}}{8}x^2y-\frac{y}{2}\sum_{m=1}^{\infty}\frac{A_{m,0}(x,y)}{m^2}+\frac{x^2}{4}\sum_{n=1}^{\infty}\frac{a_{0,n}\sin ny-c_{0,n}\cos ny}{n}$$
$$+\sum_{m,n=1}^{\infty}\Big(\frac{-a_{m,n}\cos mx\sin ny-b_{m,n}\sin mx\sin ny+c_{m,n}\cos mx\cos ny}{m^2n}$$
$$+\frac{d_{m,n}\sin mx\cos ny}{m^2n}\Big); \tag{5.3}$$

$$\frac{a_{0,0}}{8}xy^2+\frac{y^2}{4}\sum_{m=1}^{\infty}\frac{a_{m,0}\sin mx-b_{m,0}\cos mx}{m}-\frac{x}{2}\sum_{n=1}^{\infty}\frac{A_{0,n}(x,y)}{n^2}$$
$$+\sum_{m,n=1}^{\infty}\Big(\frac{-a_{m,n}\sin mx\cos ny+b_{m,n}\cos mx\cos ny-c_{m,n}\sin mx\sin ny}{mn^2}$$
$$+\frac{d_{m,n}\cos mx\sin ny}{mn^2}\Big); \tag{5.4}$$

$$\frac{a_{0,0}(x^2+y^2)}{16}-\frac{1}{2}\sum_{m=1}^{\infty}\frac{a_{m,0}\cos mx+b_{m,0}\sin mx}{m^2}$$

$$-\frac{1}{2}\sum_{n=1}^{\infty}\frac{a_{0,n}\cos ny+c_{0,n}\sin ny}{n^2}-\sum_{m,n=1}^{\infty}\frac{A_{m,n}(x,y)}{m^2+n^2}. \tag{5.5}$$

If (5.1) is the Fourier series of the function $f(x,y)\in L_2(Q)$, then in view of the inequality

$$\left|\frac{a}{m}\right|\le\frac{1}{2}\Big(a^2+\frac{1}{m^2}\Big),$$

all the above-mentioned series (5.2), (5.3), (5.4) and (5.5) converge absolutely and uniformly, and hence they define continuous functions which we denote, respectively, by $\phi_{22}(x,y)$, $\phi_{21}(x,y)$, $\phi_{12}(x,y)$ and $\phi(x,y)$.

Assume

$$\begin{gathered}\Delta^{22}(F;x,y,2u,2v)=F(x+2u,y+2v)+F(x+2u,y-2v)\\+F(x-2u,y+2v)+F(x-2u,y-2v)+4F(x,y)-2F(x+2u,y)\\-2F(x-2u,y)-2F(x,y+2v)-2F(x,y-2v),\\\Delta^{21}(F;x,y,2u,v)=F(x+2u,y+v)+F(x-2u,y+v)\\+2F(x,y-v)-2F(x,y+v)-F(x+2u,y-v)-F(x-2u,y-v),\\\Delta^{12}(F;x,y,u,2v)=F(x+u,y+2v)+F(x+u,y-2v)\\+2F(x-u,y)-2F(x+u,y)-F(x-u,y+2v)-F(x-u,y-2v).\end{gathered}$$

Lemma 5.5.2. *The equalities* ([7])

$$\begin{gathered}R_{22}(x,y,u,v)=\frac{\Delta^{22}(\phi_{22};x,y,2u,2v)}{16u^2v^2},\\R_{21}(x,y,u,v)=\frac{\Delta^{21}(\phi_{21};x,y,2u,v)}{8u^2v},\\R_{12}(x,y,u,v)=\frac{\Delta^{12}(\phi_{12};x,y,u,2v)}{8uv^2},\\\Delta\phi(x,y)=S[f]\end{gathered}$$

are valid.

Let (5.1) be the Fourier series of the continuous function $F(x,y)$. Consider the following series:

$$\sum_{m,n=0}^{\infty}\lambda_{m,n}\frac{\partial}{\partial x}A_{m,n}(x,y), \tag{5.6}$$

$$\sum_{m,n=0}^{\infty}\lambda_{m,n}\frac{\partial}{\partial y}A_{m,n}(x,y), \tag{5.7}$$

$$\sum_{m,n=0}^{\infty}\lambda_{m,n}\frac{\partial^2}{\partial x\partial y}A_{m,n}(x,y), \tag{5.8}$$

$$\sum_{m,n=0}^{\infty} \lambda_{m,n} \Delta A_{m,n}(x,y). \tag{5.9}$$

Of the terms of these series we compose the series

$$F_{22}(x,y) = \sum_{m,n=1}^{\infty} \Big(\frac{a_{m,n} \sin mx \sin ny - b_{m,n} \cos mx \sin ny - c_{m,n} \sin mx \cos ny}{mn}$$
$$+ \frac{d_{m,n} \cos mx \cos ny}{mn} \Big),$$
$$F_{21}(x,y) = \frac{y}{2} \sum_{m=1}^{\infty} \frac{a_{m,0} \sin mx - b_{m,0} \cos mx}{m}$$
$$+ \sum_{m,n=1}^{\infty} \Big(\frac{a_{m,n} \sin mx \sin ny - b_{m,n} \cos mx \sin ny - c_{m,n} \sin mx \cos ny}{mn}$$
$$+ \frac{d_{m,n} \cos mx \cos ny}{mn} \Big),$$
$$F_{12}(x,y) = \frac{x}{2} \sum_{n=1}^{\infty} \frac{a_{0,n} \sin ny - c_{0,n} \cos ny}{n}$$
$$+ \sum_{m,n=1}^{\infty} \Big(\frac{a_{m,n} \sin mx \sin ny - b_{m,n} \cos mx \sin ny - c_{m,n} \sin mx \cos ny}{mn}$$
$$+ \frac{d_{m,n} \cos mx \cos ny}{mn} \Big).$$

The functions $F_{22}(x,y)$, $F_{21}(x,y)$ and $F_{12}(x,y)$ are continuous, and ([12], p. 14)

$$F_{22}(x,y) = \int_0^x \int_0^y F(t,\tau)\,dt\,d\tau + y\varphi(x) + x\psi(y),$$
$$F_{21}(x,y) = \int_0^x \int_0^y F(t,\tau)\,dt\,d\tau + x\psi_1(y),$$
$$F_{12}(x,y) = \int_0^x \int_0^y F(t,\tau)\,dt\,d\tau + y\varphi_1(x).$$

Lemma 5.5.3. *The following equalities are valid:*

$$\Delta^{22}(F_{22}; x, y, 2u, 2v) = \int_0^{2u} \int_0^{2v} [F(x+t, y+\tau)$$

$$-F(x-t,y+\tau)-F(x+t,y-\tau)+F(x-t,y-\tau)]dtd\tau$$
$$=\Delta_{22}(F;x,y,2u,2v)16u^2v^2,$$

$$\Delta^{21}(F_{21};x,y,2u,v)=\int_0^{2u}\int_0^{v}[F(x+t,y+\tau)$$
$$-F(x-t,y+\tau)+F(x+t,y-\tau)-F(x-t,y-\tau)]dtd\tau$$
$$=\Delta_{21}(F;x,y,2u,v)8u^2v,$$

$$\Delta^{12}(F_{12};x,y,u,2v)=\int_0^{u}\int_0^{2v}[F(x+t,y+\tau)$$
$$-F(x+t,y-\tau)+F(x-t,y+\tau)-F(x-t,y-\tau)]dtd\tau$$
$$=\Delta_{12}(F;x,y,u,2v)8uv^2.$$

This lemma and Theorem 5.5.1 lead immediately to

Theorem 5.5.2. *Let $f_i(x,y)$, $i=1,2,3,4$ be arbitrary measurable and everywhere finite functions on $Q=[0,2\pi;0,2\pi]$. Then there exists the continuous function $F(x,y)$, such that if (5.1) is the Fourier series, then almost everywhere on Q, we have:*

1) *The series (5.6) is summable by the method $R(2,1)$ to $f_1(x,y)$;*
2) *The series (5.7) is summable by the method $R(1,2)$ to $f_2(x,y)$;*
3) *The series (5.8) is summable by the method $R(2,2)$ to $f_3(x,y)$;*
4) *The series (5.9) is summable by the method R^* to $f_4(x,y)$.*

There are no answers to the questions ([29], p. 78) 1) whether Theorem 5.5.2 remains valid for the case, where $f(x,y)=+\infty$ or $f(x,y)=-\infty$ on a set of positive measure; 2) whether the analogous theorem is valid for the Abelian method, or for another methods of summation, as well as for ordinary convergence in Pringsheim's sense.

Bibliography

[1] Bari N. K. On the primitive functions and trigonometric series converging almost everywhere. *Mat. Sbornik*, **31(73):3** (1952), 686–702 (in Russian).

[2] Bari N. K. *Trigonometric Series.* M. 1961 (in Russian).

[3] Beytmen G., and Erdein A. *Higher transcendent functions. Hypergeometric function, Legendre functions.* M.: Nauka, (1965) (in Russian).

[4] Beytmen G., and Erdein A. *Higher transcendent functions.* II. M.: Nauka, (1974) (in Russian).

[5] Berens H., Butzer P. L., and Pawelke S. Limitierungsverfahren von Reihen mehrdimensionaler Kugelfunktionen und deren Saturationsverhalten. *Publ. Res. Inst. Math. Sci., A*, **4** (1968), 201–268.

[6] Vilenkin N. Ja. *Special Functions and the Theory of Group Representation.* M.: Nauka (1965) (in Russian).

[7] Gakharia K. K. Summation of double trigonometric series by means of the method due to Riemann. *Mat. Sb.*, **28(70):2** (1951), 337–350 (in Russian).

[8] Germaier Ju. B. The Riemann and Valée-Poussin derivatives and their application to some problems of the theory of trigonometric series. *M.: Dissertation* (1946) (in Russian).

[9] Gobson E. V. *The Theory of Spherical and Elliptic Functions.* M.: Inostr. Literat. (1952) (in Russian).

[10] Gradshteyn I. S., and Ryjik I. M. *Tables of integrals, sums, series and products.* M. (1963) (in Russian).

[11] Jvarsheishvili A. On N. N. Luzin's theorem for a function of two variables. *Soobshch. Akad. Nauk Gruz. SSR*, XIV, **1** (1953), 11–15 (in Russian).

[12] Jvarsheishvili A. On the summation of doubl trigonometric series by Riemann's method. *Trudy Tbiliss. Mat. Inst. Razmadze Akad. Nauk Gruz. SSR*, **XX** (1954), 157–166 (in Russian).

[13] Jvarsheishvili A. On functions, analytical in a half-plane. *Trudy Tbiliss. Mat. Inst. Razmadze Akad. Nauk Gruz. SSR*, **XXXI** (1966), 91–109 (in Russian).

[14] Jvarsheishvili A. On a total differential. *Trudy Tbiliss. Mat. Inst. Razmadze Akad Nauk Gruz. SSR*, **XXXIV** (1968), 13–21 (in Russian).

[15] Jackson D. *The Fourier series and orthogonal polynomials.* M.: Inostr. Literat. (1948) (in Russian).

[16] Dzagnidze O. P. Representation of measurable functions of two variables by double series. *Soobshch. Akad. Nauk Gruz. SSR*, **34:2** (1964), 271–282 (in Russian).

[17] Dzagnidze O. P. On the boundary behavior of functions given in a sphere. *Soobshch. Akad. Nauk Gruz. SSR*, **55**]] (1969), 282–284 (in Russian).

[18] Dzagnidze O. P. On some boundary properties of functions, harmonic in a sphere. *Dokl. Akad. Nauk SSSR*, **198** (1971), 1005–1006 (in Russian).

[19] Dzagnidze O. P. On some boundary properties of functions, harmonic in a sphere. *Trudy Tbiliss. Mat. Inst. Razmadze Akad. Nauk Gruz. SSR*, **42** (1972), 65–77 (in Russian).

[20] Dzagnidze O. P. Formulas of a mixed derivative of the Poisson integral, and its boundary properties. *Soobshch. Akad. Nauk Gruz. SSR*, No 2, **120** (1985), 1–4 (in Russian).

[21] Dzagnidze O. P. Partial derivatives with a boundary behavior, and variation of the Poisson integral. *Trudy Tbiliss. Mat. Inst. Razmadze Akad. Nauk Gruz. SSR*, **76** (1985), 18–39 (in Russian).

[22] Dzagnidze O. P. Representation of a pair of functions by derivatives of the Poisson integral. *Soobshch. Akad. Nauk Gruz. SSR*, **122:1** (1986), 21–23 (in Russian).

[23] Dzagnidze O. P. Boundary values of derivatives of the Poisson integral for a sphere, and representation of functions of two variables. *Trudy Tbiliss. Mat. Inst. Razmadze Akad. Nauk Gruz.*, **98** (1990), 52–98 (in Russian).

[24] Dzagnidze O. P. On angular limits at sphere poles of derivatives of the Poisson integral. *Trudy Tbiliss. Mat. Inst. Razmadze Akad. Nauk Gruz.*, **98** (1990), 99–111 (in Russian).

[25] Dzagnidze O. P. Boundary properties of second order derivatives of the spherical Poisson integral. *Proc. A. Razmadze Math. Inst.* **102** (1993), 9–27.

[26] Dzagnidze O. P. On the differentiability of functions of two variables and of indefinite double integrals. *Proc. A. Razmadze Math. Inst.* **106** (1993), 7–48.

[27] Dyachkov A. M. On the summability of differentiated multiple Fourier series by Bochner–Riesz means. *Mat. Zametki*, No. 4, **36** (1984), 493–507 (in Russian).

[28] Zhak I. E. On the summability of double numerical series by Riemann's method due to Riemann. *Soobshch. Akad. Nauk Gruz. SSR*, **XIII:10** (1952), 587–593 (in Russian).

[29] Zhizhiashvili L. V. On some questions from the theory of primary and multiple trigonometric and orthogonal series. *UMN*, **28:2(170)** (1973), 65–119 (in Russian).

[30] Zhizhiashvili L. V. Topuria S. B. The Fourier-Laplace series on a sphere. In: Mat. Analiz, **15** (*Itogi Nauki i Tehn. VINITI Akad. Nauk SSSR*), M. (1977), 83–130 (in Russian).

[31] Zhizhiashvili L. V. Some questions of the theory of trigonometric Fourier series and their conjugates. *Izdat. TGU, Tbilisi*, 1993 (in Russian).

[32] Zhizhiashvili L. V. Some questions of the multi-dimensional harmonic analysis. *Izdat. TGU, Tbilisi*, 1996 (in Russian).

[33] Fatou P. Series trigonometriques et series de Taylor. *Acta Math.* **30** (1906), 335–400.

[34] Zygmund A. *Trigonometric Series.* I. M.: Mir (1965) (in Russian).

[35] Zygmund A. *Trigonometric series.* II. M.: Mir (1965) (in Russian).

[36] Jessen B. Marcinkiewicz J. and Zygmund A. Note on the differentiability of multiple integrals. *Fund. Math.* **25** (1935), 217–234.

[37] Kohn M. J. Summability R_r for double series. *Pacific J. Math.*, **69:2** (1977), 433–448.

[38] Kohn M. J. On formal integration of double trigonometric series. *Illinois J. Math.*, **22:1** (1978), 81–89.

[39] Kohn M. J. On differentiation of multiple trigonometric series. *Illinois J. Math.*, **25:2** (1981), 221–232.

[40] Konyagin S. V. On indeterminacy limits of trigonometric series. *Mat. Zametki*, **44:6** (1988), 770–783 (in Russian).

[41] Kratcer A., Franz V. Transcendent functions. *M.: Izd. Inostr. Literat.*, (1963) (in Russian).

[42] Luzin N. N., To the basic theorem of integral calculus. *Mat. Sb.*, **28** (1912), 266–294 (in Russian).

[43] Luzin N. N., An integral and trigonometric series. *M.-L.* (1951) (in Russian).

[44] Luzin N. N., *Collected Papers.* I. Moscow, Izdat. Akad. Nauk SSSR (1953), 400p (in Russian).

[45] Marcinkiewicz J., Sur les numbres derives. *Fund. Math.*, **24** (1935), 305–308.

[46] Marcinkiewicz J., and Zygmund A. On the differentiability of functions and summability of trigonometrical series. *Fund. Math.*, **26** (1936), 1–43.

[47] Marcinkiewicz J., and Zygmynd A. On the summability of double Fourier series. *Fund. Math.*, **32** (1939), 112–132.

[48] Menshov D. E. Sur la representation des fonctions mesurables par des series trigonometrieques. *Mat. Sb.* **9(51)** (1941), 667–692.

[49] Menshov D. E. On the convergence in measure of trigonometric series. *Trudy Mat. Inst. Steklov.,* **32** (1950), 3–97 (in Russian).

[50] Menshov D. E., and Ul'yanov P. L. On the problem of representation of functions by series. *Vestn. Moskovskogo Univ., Mat.-Mekh.*, **2** (1970), 30–40 (in Russian).

[51] Privalov I. I. Boundary value problems of the theory of harmonic and subharmonic functions in a space. *Mat. Sb.*, **3(45)** (1938), 3–25 (in Russian).

[52] Privalov I. I., and Kuznetsov P. I. Boundary value problems and different classes of harmonic and subharmonic functions defined in arbitrary spaces. *Mat. Sb.*, **6(48)** (1939), 345–376 (in Russian).

[53] Privalov I. I., *Boundary properties of analytic functions.* M.-L. (1950) (in Russian).

[54] Rauch H. E. Harmonic and analytic functions of several variables and the maximal theorem of Hardy and Littlewood. *Journal Canadien de Math.*, **VII:2**(1956), 171–183.

[55] Rudin W. Integral representation of continuous functions. *Trans. Amer. Math. Soc.*, **68** (1950), 278–286.

[56] Rudin W. Uniqueness theory for Laplace series. *Trans. Amer. Math. Soc.*, **68** (1950), 287–303.

[57] Saks S. *Integral Theory.* M.: Inostr. Literat. (1949) (in Russian).

[58] Sege G. *Orthogonal polynomials.* M.: Fizmatgiz (1962) (in Russian).

[59] Sobolev S. L. *Equations of Mathematical Physics.* M.: Nauka (1966) (in Russian).

[60] Solomentsev E. D. On boundary values of subharmonic functions. *Czechoslovak Mat. J.*, **8(83)** (1958) Prague, 520–536 (in Russian).

[61] Solomentsev E. D. On classes of functions, subharmonic in a half-space. *Vestn. Moskov. Univ.*, **73:5** (1959), 73–91 (in Russian).

[62] Solomentsev E. D. On the basic Green's formula for harmonic functions. *Sibirsk. Mat. Zh.* **VII:6** (1966), 1432–1434 (in Russian).

[63] Solomentsev E. D. Boundary functions representable by the Green type integrals I. *Izv. Akad. Nauk SSSR, Ser. Matem.* **31** (1967), 149–160 (in Russian).

[64] Solomentsev E. D. Harmonic functions representable by the Green type integrals II. *Izv. Akad. Nauk SSSR, Ser. Matem.* **31** (1967), 835–854 (in Russian).

[65] Stein I. *Singular Integrals and Differential Properties of Functions.* M.: Mir (1973) (in Russian).

[66] Stein I. and Weiss G. *Introduction to the Harmonic Analysis on the Euclidean Spaces.* M.: Mir (1974) (in Russian).

[67] Shapiro V. I. Circular summability C of double trigonometric series. *Trans. Amer. Math. Soc.*, **76:2** (1984), 223–233.

[68] Shapiro V. I. The symmetric derivative on the $k-1$-dimensional hypersphere. *Trans. Amer. Math. Soc.* **81** (1956), 514–524.

[69] Siegel A. Summability C of series of surface spherical harmonics. *Trans. Amer. Math. Soc.*, **104:2** (1962), 284–307.

[70] Talalyan A. A. Representation of measurable functions by series. *UMN*, **15:5** (1960), 77–141 (in Russian).

[71] Talalyan A. A. The problems of representation and uniqueness in the theory of orthogonal series. *Mat. Analiz,* (1970), Moscow: Nauka, (1971), 5–64 (in Russian).

[72] Tevzadze N. R. On the uniqueness of double trigonometric series. *Trudy Tbiliss. Gos. Univ.* **84** (1961), 185–192 (in Russian).

[73] Tevzadze N. R. The note on the R_λ-summability of double Fourier–Lebesgue series. *Trudy Tbiliss. Gos. Univ.* **117** (1966), 163–171 (in Russian).

[74] Timan A. F. and Trofimov V. N. *Introduction to the Theory of Harmonic Functions.* M.: Nauka (1968) (in Russian).

[75] Tolstov G. P. Remark to the D. F. Yegorov's theorem. *DAN SSSR*, **22** (1939), 309–311 (in Russian).

[76] Tolstov G. P. On the curvilinear and iterated integral. *Trudy Mat. Inst. Steklov*, **35** (1950), 1–100 (in Russian).

[77] Topuria S. B. Boundary properties of harmonic functions in the unit sphere. *Soobshch. Akad. Nauk Gruz. SSR*, **XLI:2** (1966), 265–272 (in Russian).

[78] Topuria S. B. On the existence of angular boundary values of harmonic functions in a ball. *Soobshch. Akad. Nauk Gruz. SSR*, **XLII:3** (1966), 521–527 (in Russian).

[79] Topuria S. B. Summation by Abel's method of the differentiated Fourier-Laplace series. *Soobshch. Akad. Nauk Gruz. SSR*, **54:1** (1969), 29–32 (in Russian).

[80] Topuria S. B. A general solution of the Dirichlet problem for a ball. *Soobshch. Akad. Nauk Gruz. SSR*, **55:1** (1969), 25–28 (in Russian).

[81] Topuria S. B. Classes of functions and the Fourier-Laplace series. *Trudy Gruz. Politech. Inst.*, **5(133)** (1969), 19–32 (in Russian).

[82] Topuria S. B. On local boundary properties of harmonic functions in a ball. *Trudy Gruz. Politech. Inst.*, **5(133)** (1969), 33–37 (in Russian).

[83] Topuria S. B. Solution of the Dirichlet problem for a half-space. *Dokl. Akad. Nauk SSSR*, **195:3** (1970), 567–569 (in Russian).

[84] Topuria S. B. On one property of functions of two variables and multiple Fourier series. *Dokl. Akad. Nauk SSSR*, **195:5** (1970), 1046–1049 (in Russian).

[85] Topuria S. B. On the representation of functions defined on the surface of a unit sphere by singular integrals, and summability of Laplace series. *Trudy Gruz. Politech. Inst.*, **7(147)** (1971), 25–53 (in Russian).

[86] Topurua S. B. Summation by Abel's method of the differentiated Fourier-Laplace series. II. *Trudy Gruz. Politech. Inst.*, **7(147)** (1971), 17–24 (in Russian).

[87] Topuria S. B. Summation by Abel's method of the differentiated Fourier series. *Dokl. Akad. Nauk SSSR*, **209:3** (1973), 569–572 (in Russian).

[88] Topuria S. B. On one method of summation of double Fourier series. *Soobshch. Akad. Nauk Gruz. SSR*, **76:2** (1974), 289–292 (in Russian).

[89] Topuria S. B. Boundary properties of the differentiated Poisson integral in a circle. *Trudy Gruz. Politech. Inst.*, **176:3** (1975), 13–20 (in Russian).

[90] Topuria S. B. The boundary properties of the differentiated Poisson integral in a bicylinder. I. *Trudy Gruz. Politech. Inst.*, **176:3** (1975), 21–32 (in Russian).

[91] Topuria S. B. The boundary properties of the differentiated Poisson integral in a half-plane. *Soobshch. Akad. Nauk Gruz. SSR*, **78:2** (1975), 281–284 (in Russian).

[92] Topuria S. B. On the representation of functions defined on a surface of unit sphere by Laplace series. *Soobshch. Akad Nauk Gruz. SSR*, **82:2** (1976), 301–303 (in Russian).

[93] Topuria S. B. The boundary properties of the differentiated Poisson integral, and solution of the Dirichlet problem in a half-space. *Trudy Gruz. Politech. Inst.*, **197:6** (1977), 11–30 (in Russian).

[94] Topuria S. B. On the representation of functions of several variables by trigonometric series. *Trudy Gruz. Politech. Inst.*, **5(237)** (1981), 9–11 (in Russian).

[95] Topuria S. B. *The Fourier–Laplace Series on a Sphere.* Tbilisi: Izdat. Tbiliss. Gos. Univ. (1987) (in Russian).

[96] Topuria S. B. The boundary properties of the differentiated Poisson integral in a half-space, and representation of a function of two variables. *Soobshch. Akad. Nauk Gruz. SSR*, **152:2** (1995), 235–238 (in Russian).

[97] Topuria S. B. The boundary properties of derivatives of the Poisson integral in a half-space. *Soobshch. Akad. Nauk Gruz. SSR*, **152:3** (1995), 453–457 (in Russian).

[98] Topuria S. B. Generalized derivatives and boundary properties of differentiated Poisson integral for a disk and a half-plane. *Bull. Georgian Acad. Sci.*, **153:3** (1996), 329–332.

[99] Topuria S. B. The boundary properties of derivatives of the Poisson integral in a half-space, and representation of a function of two variables. *Dokl. Akad. Nauk*, **353:3** (1997), 306–309 (in Russian).

[100] Topuria S. B. Generalized Laplacian and boundary properties of differentiated Poisson integral for a half-space. *Bull. Georgian Acad. Sci.*, **155:2** (1997), 164–167.

[101] Topuria S. B. Generalized partial derivatives of functions of many variables and boundary value problems. *Bull. Georgian Acad. Sci.*, **155:3** (1997), 324–327.

[102] Topuria S. B. Generalized total differentials of many variable functions and boundary value problems. *Bull. Georgian Acad. Sci.*, **156:1** (1997), 22–26.

[103] Topuria S. B. Generalized spherical derivatives and boundary value problems. *Bull. Georgian Acad. Sci.*, **156:2** (1997), 181–185.

[104] Topuria S. B. Boundary properties of first order partial derivatives of the Poisson integral for the half-space R_+^{k+1} $(k > 1)$. *Georgian Math. J.*, **4:6** (1997), 585–600.

[105] Topuria S. B. Boundary properties of second order partial derivatives of the Poisson integral for the half-space R_+^{k+1} $(k > 1)$. *Georgian Math. J.*, **5:4** (1998), 385–400.

[106] Topuria S. B. The generalized Laplace operator on a sphere and summation of differentiated Fourier-Laplace series using the Abel's method. *Bull. Georgian Acad. Sci.*, **160:3** (1999), 409–412.

[107] Topuria S. B. Generalized derivatives and boundary properties of derivatives of the Poisson integral for the half-space R_+^{k+1} $(k > 1)$. *Analysis Mathematica*, **25:1** (1999), 69–86 (in Russian).

[108] Topuria S. B. Generalized derivatives of an arbitrary order and the boundary properties of differentiated Poisson integral for the half-space R_+^{k+1} $(k > 1)$. *Georgian Math. J.*, **7:2** (2000), 387–400.

[109] Topuria S. B. The generalized Laplace operators in the space R^k $(k \geq 2)$ and on the unit sphere. *Trudy Gruz. Techn. Univ.*, **430:2** (2000), 9–18 (in Russian).

[110] Topuria S. B. The generalized Laplace operators on the unit sphere. *Bull. Georgian Acad. Sci.*, **173:1** (2006), 11–13.

[111] Topuria S. B. The boundary properties of the differentiated conjugate Poisson and Poisson-Stiltjes integral. *Trudy Gruz. Tekh. Univ.*, **460:2** (2006), 92–95 (in Russian).

[112] Witteker E. T. and Vatson J. N. *The Course of the Modern Analysis. Part* 2. *Transcendent functions.* M.: Fizmatgiz (1963) (in Russian).

[113] Ul'yanov P. L. Solved an unsolved problems of the theory of trigonometric and orthogonal series. *UMN*, **19:1** (1964), 3–69 (in Russian).

[114] Ul'yanov P. L. Representation of functions by series and the classes $\varphi(L)$. *UMN*, **27:2(164)** (1972), 3–52 (in Russian).

[115] Fichtenholz G. M. The course of differential and integral calculus. *M.-L. Fizmatgiz*, **III** (1960) (in Russian).

[116] Ward A. J. On the derivation of additive functions of intervals in m-dimensional space. *Fund. Math.* **28** (1937), 265–279.

[117] Wheeden R. L. Derivatives of Fourier series and integrals. *Stud. Math.*, **XXVIII** (1966), 69–80.

[118] Housdorff F. *The Theory of Sets.* TTL (1937) (in Russian).

Index